T0259690

SpringerBriefs in Applied Sciences and Technology

SpringerBriefs present concise summaries of cutting-edge research and practical applications across a wide spectrum of fields. Featuring compact volumes of 50–125 pages, the series covers a range of content from professional to academic.

Typical publications can be:

- A timely report of state-of-the art methods
- An introduction to or a manual for the application of mathematical or computer techniques
- A bridge between new research results, as published in journal articles
- A snapshot of a hot or emerging topic
- An in-depth case study
- A presentation of core concepts that students must understand in order to make independent contributions

SpringerBriefs are characterized by fast, global electronic dissemination, standard publishing contracts, standardized manuscript preparation and formatting guidelines, and expedited production schedules.

On the one hand, **SpringerBriefs in Applied Sciences and Technology** are devoted to the publication of fundamentals and applications within the different classical engineering disciplines as well as in interdisciplinary fields that recently emerged between these areas. On the other hand, as the boundary separating fundamental research and applied technology is more and more dissolving, this series is particularly open to trans-disciplinary topics between fundamental science and engineering.

Indexed by EI-Compendex, SCOPUS and Springerlink.

More information about this series at http://www.springer.com/series/8884

Russell Wanhill · Simon Barter ·
Loris Molent

Fatigue Crack Growth Failure and Lifing Analyses for Metallic Aircraft Structures and Components

 Springer

Russell Wanhill
Emmeloord, The Netherlands

Loris Molent
Aerospace Division
DST Group
Melbourne, Australia

Simon Barter
Aerospace Division
DST Group
Melbourne, Australia

ISSN 2191-530X   ISSN 2191-5318   (electronic)
SpringerBriefs in Applied Sciences and Technology
ISBN 978-94-024-1673-2   ISBN 978-94-024-1675-6   (eBook)
https://doi.org/10.1007/978-94-024-1675-6

Library of Congress Control Number: 2018966847

This Springer imprint is published by the registered company Springer Nature B.V.
The registered company address is: Van Godewijckstraat 30, 3311 GX Dordrecht, The Netherlands

Preface

This book provides a concise discussion of fatigue crack growth (FCG) failure and lifing analysis methods for metallic aircraft structures and components. After a reasonably concise historical review, surveys are made of (i) the importance of fatigue for aircraft structural failures and the sources of fatigue nucleation and cracking, (ii) contemporary FCG lifing methods, and (iii) the quantitative fractography (QF) required for determining the actual FCG behaviour. These surveys are followed by the main part of the book, which is a discussion, using case histories, of the applicabilities of linear elastic fracture mechanics (LEFM) and non-LEFM methods for analysing service fatigue failures and full- and sub-scale test results. This discussion is derived primarily from the experiences of the Defence Science and Technology Group in Melbourne, Australia, and the Netherlands Aerospace Centre, Marknesse, the Netherlands.

The opinions expressed in this book are those of the authors and do not necessarily represent those of the organisations with which they are, or have been, associated.

Emmeloord, The Netherlands   Russell Wanhill
Melbourne, Australia   Simon Barter
Melbourne, Australia   Loris Molent

Contents

Abstract

This book considers fatigue crack growth (FCG) failure and lifing analysis methods for metallic aircraft structures and components, based on the experiences of the Australian Defence Science and Technology Group and the Netherlands Aerospace Centre. A reasonably concise historical review is followed by surveys of basic information for aircraft FCG failure analyses and the quantitative fractography (QF) required for determining the actual FCG behaviour. These surveys are followed by a selection of case histories and lifing analyses.

Keywords Metallic aircraft structures · Fatigue crack growth · Failure analysis · Quantitative fractography · Lifing of aircraft structures

Chapter 1
Historical Review

Fatigue crack growth (FCG) analyses for metals and alloys properly began in the 1950s [1–6]. Test data [3–6] were obtained by optical measurements of the crack lengths visible on the specimen surfaces. In the same decade it was suspected [7] and demonstrated by optical fractography [8] that striations on fatigue fracture surfaces generally represented cycle-by-cycle progression of a crack. This demonstration was achieved by using simple block programme loading, and may be considered as the origin of Quantitative Fractography (QF) for FCG analyses.

A vast number of publications (papers, reports and books) on FCG have been issued over the intervening decades. The review in this Chapter is directed to partic-ularly relevant topics for FCG failure and lifing analyses.

1.1 FCG Parameters, Concepts and Testing

The most important development in the early 1960s was the introduction by Paris et al. of the linear elastic fracture mechanics (LEFM) stress intensity factor K_{max} to describe and correlate constant amplitude (CA) and constant stress ratio FCG data [9]. Subsequently, Paris and Erdogan [10] used the stress intensity factor range ΔK, and this has since become the standard basic parameter for correlating FCG data. In the late 1960s Elber discovered plasticity-induced fatigue crack closure [11, 12] leading to defining ΔK_{eff}, the effective ΔK for *long cracks*. At the same time the concept of the FCG threshold, ΔK_{th}, was first mentioned explicitly by Hartman and Schijve [13]. An extensive review and analysis of FCG thresholds for long cracks was published by Chan in 2004 [14], from which it is evident that this is a complex topic.

In 1968 the well-known Rainflow cycle counting method for variable amplitude (VA) loading was introduced [15]. This method has become widely accepted [16], and a cycle-counting algorithm is included in ASTM Standard E1049–85 [17]. Rainflow counting may be used in estimating fatigue lives, as originally intended, *and* FCG lives.

R. Wanhill et al., *Fatigue Crack Growth Failure and Lifing Analyses for Metallic Aircraft Structures and Components*, SpringerBriefs in Applied Sciences and Technology, https://doi.org/10.1007/978-94-024-1675-6_1

In the 1970s several important developments were begun:

(1) The use of characteristic K-values to (attempt to) correlate FCG data obtained under VA loading [18–20]. Since then this approach has been developed further, notably for combat aircraft, e.g. [21, 22].

(2) The United States Air Force (USAF) introduced mandatory *safety* and *durability* guidelines to ensure aircraft structural integrity [23, 24] after the crash of a General Dynamics F-111A aircraft in December 1969 and early and widespread fatigue cracking of Lockheed C-5A wing boxes [25]. These guidelines became known as the 'Damage Tolerance philosophy'. A key feature was that initial damage (cracks or crack-like flaws) in *new* structures had to be assumed.

     The assumed initial crack/flaw sizes to ensure *safety* were large enough for LEFM-based FCG analyses to use macrocrack test data. However, the assumed initial crack/flaw sizes for *durability* analyses (intended to ensure an economic FCG life) were much smaller, well below 0.5 mm. These crack/flaw sizes were in what is now commonly referred to as the *short crack regime*, making the durability analyses problematical, see point (3).

(3) In the mid-to-late 1970s the apparently anomalous FCG behaviour of short cracks was discovered and investigated [26–28]. Research on this topic has since expanded enormously, with several landmark volumes already in the 1983–1990 time period [29–32]. Information in these volumes showed that short crack FCG can be influenced by many factors, e.g. [33, 34], including the partial or total absence of crack closure [33]. In turn, this information suggested that the USAF's original method of obtaining *durability* FCG lives, which included back-extrapolation of macrocrack test data, was unreliable [35].

     In the intervening decades it has become clear that analytical modelling of short crack FCG properly requires validation by QF-obtained data, see Chap. 2, Sect. 2.2.2. It is also important to note that such analyses are now regarded as relevant to linking *safety* and *durability* for ageing aircraft. **N.B:** the term *sustainment* has become preferred to the general use of *durability*, in recognition that fatigue is not the only issue in estimating the useful life of ageing aircraft, e.g. [36].

(4) An ASTM standard method of FCG testing. The current edition is ASTM E647–15e1 [37]. The standard was first proposed in 1978 [38] and issued as E647–78T, with T signifying tentative. This 'T' designation turned out to be well justified: over the intervening years there have been numerous revisions to the standard, reflecting the importance of crack closure and especially the difficulties in determining reliable near-threshold FCG data and threshold (ΔK_{th}) values.

1.2 FCG 'Laws' and Models

A convenient background to discussing FCG 'laws' is the well-known double-logarithmic FCG rate (da/dN) versus ΔK curve, shown schematically in Fig. 1.1.

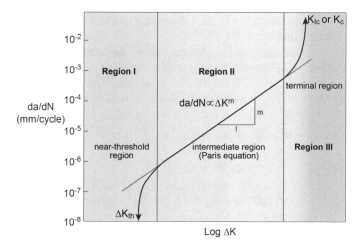

Fig. 1.1 Schematic da/dN versus ΔK diagram for long crack growth under CA loading in normal air, showing three regions of crack growth. ΔK_{th} is the FCG threshold; K_{Ic} and K_c are the plane strain and non-plane strain fracture toughnesses, respectively

The curve shape is reasonably generic for long crack growth under CA loading in a normal air environment. There are three main regions of crack growth, according to the curve shape, the mechanisms of crack growth and various influences on the curve, as follows [39, 40]:

Region I: Non-continuum crack growth mechanisms with large influences of microstructure, mean stress and environment (if different from normal air).

Region II: Crack growth commonly characterised by fatigue striations, and with small-to large influences of microstructure, depending on the material; also large effects of certain combinations of environments, mean stress and cycle frequencies.

Region III: Large contributions of 'static mode' mechanisms, generally microvoid coalescence, and in some materials cleavage or intergranular fracture; and little influence of the environment.

1.2.1 Long Crack Growth 'Laws'

FCG relations ('laws') were already being proposed in the 1950s. Hoeppner and Krupp [41] list some 16–17 relations preceding the Paris and Erdogan 1963 paper [10]. This paper is important for three main reasons: (i) it introduced ΔK as the basic parameter for correlating FCG data, as stated in Sect. 1.1; (ii) a review of the more interesting antecedent [2, 3] and contemporary [42] relations; and (iii) a comparison of these relations with Paris' and Erdogan's own da/dN versus ΔK 'law'.

This comparison concerned the Region II exponent m, see Fig. 1.1, for a standard aerospace aluminium alloy, AA2024-T3.

Following Paris and Erdogan [10], but not exactly, the more interesting antecedent and contemporary relations may be expressed as follows:

Head [2] $\quad\quad\quad\quad\quad\quad$ $da/dN = C_1\sigma^2 a/(C_2-\sigma) = C_1 K^2/(C_2-\sigma)$ $\quad\quad$ (1.1)

Frost and Dugdale [3] $\quad$ $da/dN = C_3\sigma^3 a = C_3 K^2\sigma$ $\quad\quad\quad\quad\quad\quad$ (1.2)

Liu [42] $\quad\quad\quad\quad\quad\quad$ $da/dN = C_4(\Delta\sigma)^2 a$ $\quad\quad\quad\quad\quad\quad\quad$ (1.3)

where C_1, C_2, C_3 and C_4 are constants; $\sigma =$ stress; and a = crack length. **N.B**: Eq. (1.3) could have been written as $da/dN = C_4(\Delta K)^2$, in which case Liu [42] would have *explicitly* anticipated the use of ΔK by Paris and Erdogan [10]. They proposed that the best fit to the 2024-T3 FCG data was given by

$$da/dN = C_5(\Delta K)^4 \quad\quad\quad\quad (1.4)$$

Equation (1.4) is purely empirical. Hence the value of the exponent (m = 4) is not immutable, as was quickly commented upon [43]. Also, it is evident from Eqs. (1.1–1.4) that the K or ΔK exponents depend on the selected relation, e.g. m = 2 for Eqs. (1.2 and 1.3) and m = 4 for Eq. (1.4). Thus the relation exemplified by Eq. (1.4) soon became generically written as

$$da/dN = C(\Delta K)^m \quad\quad\quad\quad (1.5)$$

This has become known as the Paris equation or 'law', which has often been used to describe the double-logarithmic approximately-linear relationship in Region II.

Since the early 1960s the search for more comprehensive and accurate FCG 'laws' has been unceasing. By 1974 Hoeppner and Krupp [40] listed 33 relations, including those not based on the use of ΔK; and by 2009 Hoeppner had collated at least 186 relations [44]. Most are based only on CA loading, often using the Paris equation as the starting point. Two of the earlier well-known modified Paris equations are those of Forman et al. [45] and Hartman and Schijve [13]:

Forman et al.[45] $\quad\quad$ $da/dN = \dfrac{C(\Delta K)^m}{(1-R)K_c - \Delta K}$ $\quad\quad\quad$ (1.6)

Hartman and Schijve [13] $\quad$ $da/dN = \dfrac{C(\Delta K - \Delta K_{th})^m}{(1-R)K_c - \Delta K}$ $\quad\quad$ (1.7)

where K_c is the fracture toughness (often an empirical fit). The denominator, $(1-R)K_c-\Delta K$, was introduced to account for the effect of stress ratio $(R = \sigma_{min}/\sigma_{max})$ on FCG rates and also to describe crack growth in Region III [45]. Subsequently ΔK_{th}

was introduced into the numerator to extend Region II linear behaviour into Region I [13].

Further modifications have been gradually added to (i) account for fatigue crack closure [22, 46]; (ii) the R-dependence of the transition points between Regions I and II, and between Regions II and III [47]; (iii) obtain improved empirical estimates of ΔK_{th} [47]; and (iv) modify the long crack FCG relations to take account of short crack FCG characteristics, thereby potentially extending the applicability of long crack FCG relations to short cracks [22, 47].

These modifications and the attendant increasing complexities have been motivated by the intention of using long crack CA test data to obtain reliable predictions of FCG in components and structures, particularly under VA loading representative of service use. An example is the Forman-Newman-de Koning (FNK) equation [48, 49] used in early versions of the NASGRO suite of computer modelling programs:

$$\text{FNK equation [48, 49]} \quad da/dN = C \left(\frac{1 - f}{1 - R} \right)^m (\Delta K)^{m-p} K_c^q \frac{(\Delta K - \Delta K_{th})^p}{(K_c - K_{max})^q} \quad (1.8)$$

where f is an empirical crack-opening function accounting for the presence or absence of crack closure; K_{max} is the maximum stress intensity factor; and p and q are empirical constants. **N.B**: In early NASGRO versions, designated NASA/FLAGRO [48, 49], the crack opening function f was already a complicated parameter [46]. Later versions tend to include yet more complexity, e.g. modifications (iii) and (iv) mentioned above.

1.2.2 Long Crack Growth Models: LEFM Analyses

The starting point for these models and analyses is the fitting of Paris-type equations, as exemplified in Sect. 1.2.1, to long crack CA test data. Additional important (essential) features are the use of (i) appropriate geometry correction factors, β, for real components and structures; (ii) a cycle counting algorithm, e.g. Rainflow, to obtain cycle-by-cycle load sequences representative of service load histories and/or realistic load histories during full-scale and component tests. These added features mean that all LEFM-based FCG prediction models estimate crack growth via a cycle-by-cycle (or positive half-cycle) sequence using numerical integration algorithms [50, 51].

Figure 1.2 surveys the types of models and some of those available. Discussions of these model types are given in Refs. [50, 52]: strip yield models are generally considered to be the most accurate [22, 52]. One of the most advanced strip yield models, FASTRAN, is illustrated by the flow chart in Fig. 1.3. Note that there is a special module to account for the crack closure behaviour of short cracks, namely the near-absence of crack closure [22]. **N.B**: Strictly speaking, short fatigue crack growth generally violates LEFM conditions, since crack tip plastic zones are usually too large with respect to the crack dimensions.

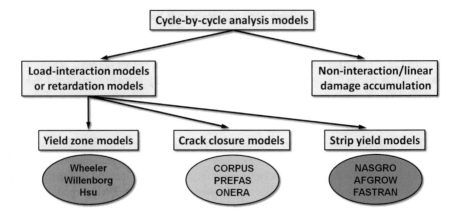

Fig. 1.2 Overview of LEFM-based FCG analysis models: after Iyyer et al. [51]. **N.B**: AFGROW is actually a framework for FCG modelling that works with several types of models, including the most advanced (strip yield) models

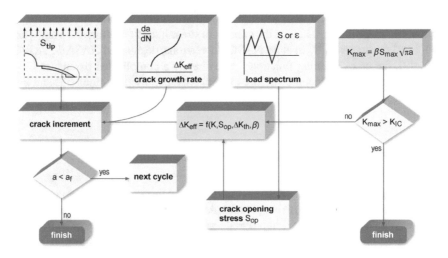

Fig. 1.3 FCG analysis flowchart [50] for the FASTRAN strip yield/crack closure model FASTRAN [53]. The module at top left is a special one for calculating the crack opening stress (and hence any crack closure) for short cracks

It is important to note that despite many years of developing LEFM-based long crack growth models, reviews by Paris et al. [54] and Ciavarella et al. [55] suggested that further improvements are needed with respect to predicting the accumulation of service FCG damage. In fact, this has been demonstrated by an extensive Round Robin programme based on realistic VA testing of (simulated) helicopter fuselage frames with rather complex crack configurations and changes during crack growth [56, 57]. The participants were 'challenged' to predict the FCG curves and lives

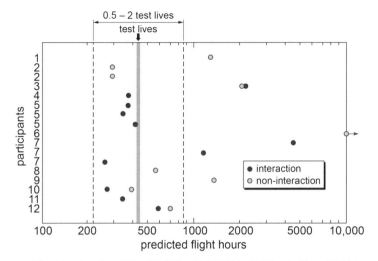

Fig. 1.4 Summary of LEFM-based long crack growth model FCG predictions compared to test results for (simulated) helicopter fuselage frames: after Irving et al. [56]

using all relevant information *except* the actual test data, i.e. these were so-called *blind (uncalibrated) predictions*. The results are summarised in Fig. 1.4, and were clearly unsatisfactory. On the other hand *calibrated* predictions, whereby models take account of the actual FCG data, can be very successful. This was demonstrated by Tiong and Jones [58] for the same helicopter fuselage panels used in the Round Robin investigation.

The issue of uncalibrated and calibrated LEFM models and analyses has been especially addressed in a subsequent review [59], where it was concluded that:

(1) Uncalibrated predictions are unsuitable for service failures, since the predictions are too unreliable. This situation is unlikely to change, owing to many uncertainties and problems. These include complex geometries; unanticipated secondary bending and residual stresses; and load-shedding, owing to loss of stiffness, from cracked to uncracked components.

(2) Calibrated models fitted to service and test data may be used—within limits—for FCG life predictions to indicate the effects of usage severity for service aircraft. The limits are that predictions should be made only for minor differences in overall stress levels (10–20%) and load histories.

In their review paper Paris et al. [54] also suggested that instead of concentrating on details (particularly referring to crack closure) the FCG modelling should be kept simple in 'solving the major problem of a reasonable yet simple accumulation of damage model'. This pragmatic suggestion is subsequent to some uses of characteristic K-values as attempts to correlate FCG data obtained under VA loading [18–20, 60–63], but precedes further developments by the DST [21, 64–66], namely:

(1) The Effective Block Approach (EBA) [21, 64].
(2) The Lead Crack Fatigue Lifing Framework (LCFLF) [65, 66].

Basic information on the EBA and LCFLF is given in Sects. 1.2.4 and 1.2.5, respectively.

1.2.3 Characteristic K Approaches

For some types of VA load histories FCG is a regular process because the larger load excursions have either a short recurrence period or only minor effects on subsequent crack growth. These VA load histories may be referred to as 'stationary' or 'quasi-stationary'. Important examples are load histories pertaining to high-strength components in tactical aircraft [62–64], in part because training missions with similar load histories are repeated at regular intervals. Quasi-stationary approximations to the service load histories of tactical aircraft are very useful for full-scale, component and specimen tests. The service load histories for transport aircraft, notably gust loads, are less amenable to such approximations; but even so, a judicious choice of test spectrum may fit into the quasi-stationary category while still giving useful test results.

The incentive for using stationary or quasi-stationary test load sequences as approximations to service use is the possibility of correlating the FCG rates by a characteristic K-value. In this approach, repeated sequences (blocks) of simulated flights are assigned a characteristic K-value as if they were single load cycles. If successful, such correlations enable efficient predictions of service FCG lives.

The first uses of characteristic K-values dates from the early 1970s [18–20]. These may be regarded as 'proof of concept' exercises. Long crack VA FCG correlations were often attempted using the root mean value of the stress intensity factor range, ΔK_{rm}:

$$\Delta K_{rm} = \sqrt[m]{\frac{\sum (\Delta K_i)^m n_i}{\sum n_i}} \tag{1.9}$$

where n_i is the number of load cycles corresponding to ΔK_i and m is the exponent in the CA Paris equation, Eq. (1.5). The exponent was usually assigned the value m = 2, resulting in the root mean square ΔK-value, ΔK_{rms}, e.g. for bridge steels [19].

In the mid-1970s to mid-1980s long crack FCG data from flight simulation tests on aluminium alloy specimens were correlated by K_{max} (tactical aircraft) [20, 63] and K_{mf} (transport aircraft) [63]: K_{max} was derived from the maximum frequently occurring stress in the tactical aircraft test spectra, and K_{mf} was derived from the so-called mean-stress-in-flight in the transport aircraft test spectrum. Better correlations were obtained (i) for tactical aircraft spectrum loading, and (ii) higher strength aluminium alloys (AA7XXX vs. AA2XXX). The reason for this latter result is that stronger alloys undergo less crack tip yielding during tensile peak loads, resulting in shorter delays in crack growth after the peak loads and hence less disturbance to a regular crack growth process.

Innovative flight simulation characteristic K correlations were developed in the mid-1970s by Gallagher and Stalnaker [20], and in the late 1980s by Wanhill [61] and De Jonge [62]:

(1) Gallagher and Stalnaker used long crack K_{max} correlations of tactical aircraft flight simulation FCG data to develop two types of normalized crack growth (NCG) curves. The first type was for scheduling modifications and structural repairs. The second type was for establishing inspection schedules and flight safety limits. The reliability of the NCG curves was checked by test data. This work resulted in the first practical extension and application of characteristic K correlations, specifically for the USAF F-4 aircraft. Subsequently, Gallagher et al. used a similar method to estimate FCG in repaired lower wings of USAF F-16 aircraft [60].

(2) Wanhill used K_{mf} to correlate Fokker 100 transport aircraft flight simulation FCG data for naturally-occurring short cracks and artificially-induced long cracks in AA2024-T3 specimens, see Fig. 1.5. The upper bound and best fit lines were then used in the following durability analysis procedure:

   • Assume the occurrence of single-hole or multiple-hole FCG (multiple site damage, MSD) in a row of fastener holes, starting from initial 0.127 mm corner cracks.
   • Generate durability (sustainment) FCG curves and predict the crack lengths at a major structural inspection life of 45,000 flights and an economic repair life of 90,000 flights.

   The best fit predictions indicated no durability problems, i.e. single or MSD cracks would be below the detection limit of standard non-destructive inspection (NDI) during the economic repair life. These predictions were consistent with full-scale test results for a Fokker 100 airframe nominally free from cracks or crack-like discontinuities. On the other hand, the upper bound predictions were too pessimistic: although no cracks would be detectable at 45,000 flights, there would be the possibility of MSD-induced *failures* along fastener rows at 90,000 flights [61]. Nevertheless, given the simplicity of the analysis and the conservative assumption of initial cracks, the overall predictions were encouraging as a 'proof of concept' exercise, namely (i) using characteristic K-values for correlating flight simulation short and long crack FCG data, and (ii) using such correlations, validated by full-scale testing, to predict in-service FCG behaviour over the entire life of an aircraft.
   N.B: The absence of an FCG threshold for short cracks, see Fig. 1.5, is most significant for durability analyses. This was also recognised by Lincoln and Melliere [35] with respect to USAF F-15E aircraft.

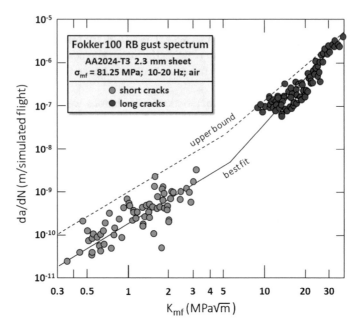

Fig. 1.5 Characteristic K correlation of short and long FCG rates for AA2024-T3 sheet tested under Fokker 100 transport aircraft flight simulation loading: after [61]

(3) De Jonge developed the crack severity index (CSI) for the Royal Netherlands Air Force (RNLAF) F-16 aircraft [62]. The starting point is a Paris-type characteristic K expression for VA (spectrum) loading,

$$da_i = C(\Delta K_{eff,i})^m = C(\beta\sqrt{\pi a})^m (\Delta\sigma_{eff,i})^m \tag{1.10}$$

representing crack growth due to load cycle i: β is the structural geometry correction factor; $\Delta\sigma_{eff,i} = (\sigma_{max,i} - \sigma_{op,i})$ or $(\sigma_{max,i} - \sigma_{min,i})$, whichever is the greater; and $\sigma_{op,i}$ is the crack opening stress for load cycle i. Eq. (1.10) may be rewritten as:

$$\frac{da_i}{(\beta\sqrt{\pi a})^m} = C(\Delta\sigma_{eff,i})^m \text{ or } f(a)da_i = C(\Delta\sigma_{eff,i})^m \tag{1.11a, b}$$

Under spectrum loading a crack of initial length a_0 will grow to a final length a_f. Defining the integral of f(a) as F(a), the amount of crack growth can be calculated from:

$$F(a_f) - F(a_0) = C\sum_{i=1}^{n} (\Delta\sigma_{eff,i})^m \tag{1.12}$$

where n is the number of load cycles corresponding to $\Delta\sigma_{eff,i}$. In Eq. (1.12) the L.H. side is defined by the crack lengths, structural geometry, and material FCG properties (via the exponent m). The R.H. side includes load spectrum terms, and defines the spectrum severity with respect to FCG *provided that* $\sigma_{op,i}$ is independent of crack length. With this proviso, and for the same material, two different spectra having the same value for the R.H. side of Eq. (1.12) will result in the same amount of FCG. This R.H. term may therefore be called the *crack severity index*, CSI [62].

To use the CSI in comparing different spectra it is necessary to calculate $\sigma_{op,i}$ for each spectrum. As explained in [62], this was done for typical tactical aircraft spectrum load sequences using the crack closure model CORPUS (included in Fig. 1.2). The calculations enabled simplifying the $\sigma_{op,i}$ variations during the load sequences, and the resulting simplified CSI approach was validated by FCG tests on specimens subjected to F-16 wing root bending spectra. The validated CSI approach was subsequently incorporated into the RNLAF Fatigue Load Monitoring Programme for the F-16 fleet [62].

1.2.4   The Effective Block Approach (EBA) Framework

The EBA [21, 64, 67, 68] is a framework whose purpose is to predict the FCG lives of in-service aircraft structures subjected to relatively short and repeated blocks of VA loading sequences, i.e. load sequences representative of tactical aircraft. The EBA has a long history, dating from a so-called mini-block proposal by Gallagher in 1976 [69]. The current EBA has been developed over the last two decades by the DST, using their extensive experience with VA full-scale fatigue testing in combination with specimen tests and QF [67, 68].

The EBA is based partly on the analysis of FCG test data obtained from VA block loading sequences simulating those experienced by tactical aircraft. Use of the EBA for predicting service-induced FCG requires several additional inputs and conditions/assumptions:

(1) Quantitative Fractography (QF) of fatigue fracture surfaces from full-scale or specimen tests (see point 2) to find FCG progression markings and measure the concomitant crack sizes. This is to obtain crack size, a, versus N_B data, where N_B is the number of applied VA blocks.
     N.B: FCG progression markings are mostly the result of judiciously chosen intermittent marker loads, or marker load combinations, distributed among the VA blocks and designed to avoid significant disturbance of the overall crack growth process [70]. However, sometimes a VA loading sequence can provide 'natural' markers.
(2) The FCG test data must be for the same material tested with a VA block loading history (as many repeat blocks as possible) considered to be representative of in-service usage.

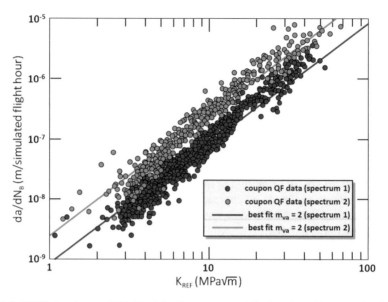

Fig. 1.6 FCGR data for two EBA block loading spectra and the best fits with $m_{va} = 2$. The QF data were obtained from many AA7050-T7451 coupons tested at four σ_{REF} levels per spectrum: after [21, 68]

(3) Each block of VA loading may be treated as if it were a single CA cycle, and crack growth associated with each block should correlate well with a similitude parameter, e.g. an appropriate (characteristic) K-value.
(4) Either specimen or full-scale fatigue test QF data (or both) may be used, but only one source of data is actually required.
(5) Conversion and presentation of the QF FCG data according to a da/dN_B versus characteristic K relationship:

$$da/dN_B = C_{va}(K_{REF})^{m_{va}} \qquad (1.13)$$

where C_{va} and m_{va} are analogous to the constant and exponent in Eq. (1.5); and K_{REF} is the reference (characteristic) K-value given by $K_{REF} = \sigma_{REF}\beta\sqrt{\pi a}$, where σ_{REF} is a reference stress (e.g. peak stress) from the VA block loading sequence.

In practice, C_{va} is determined via use of an additional LEFM-based FCG model, for example AFGROW, to determine C_{va}. However, it is recommended to set $m_{va} = 2$, based on the results of many EBA tests. A representative example is given in Fig. 1.6, which shows converted QF test data, i.e. fatigue crack growth rates (FCGR), plotted against K_{REF} using log–log coordinates. It is seen that $m_{va} = 2$ provides good fits to the data. **N.B**: This result is highly significant, see Sect. 1.2.5, since an exponent of 2 corresponds to exponential FCG, i.e. a straight line when the log crack size (depth) is plotted against linear life N (in this case N_B).

More details about the EBA, including FCG life predictions, are given especially by McDonald [64] and also in other publications, e.g. [21, 67, 68]. An example of using the EBA for service life predictions is given in Chap. 6.

1.2.5 Exponential FCG Behaviour

FCG and FGCR are sometimes described by the following general equations:

$$\textbf{FCG}: \quad a = a_0 \exp(\lambda N) \qquad (1.14)$$

$$\textbf{FCGR}: \quad da/dN = A \exp(Ba) \qquad (1.15)$$

where a_0 is the initial crack size; λ is a constant that includes β, the structural geometry correction factor; and A and B are constants. **N.B**: all three constants depend on the load history and material. Also, N here represents the FCG life as (i) cycles for CA loading; (ii) load blocks or simulated flight hours during full-scale and VA specimen testing; and (iii) flight hours for service experience.

Equations (1.14) and (1.15), and specific versions of them, are commonly referred to as representing 'exponential FCG behaviour'. They are important, especially Eq. (1.14), for FCG life estimations, since they can be *directly* applicable to analysing service FCG problems as well as full-scale and specimen test results. Three additional points need mentioning before discussing these relationships:

(1) Quantitative Fractography (QF) is required to validate Eqs. (1.14) and (1.15) for service and full-scale test cracking, and also for early crack growth in specimens.
(2) Exponential FCG equations can be appropriate to both CA [3] and VA load histories. Their main applications are in VA FCG analyses for tactical aircraft and the LCFLF, which is discussed in Sect. 1.2.5.1.
(3) Exponential FCGR equations have been found useful for *effectively* CA loading [71–73] as well as the VA block loading sequences in the EBA when $m_{va} = 2$ in Eq. (1.13).

1.2.5.1 Exponential FCG: Lead Crack Fatigue Lifing Framework (LCFLF)

The first use of Eq. (1.14) was by Frost and Dugdale [3] in 1957 to describe specimen surface observations of FCG under CA loading. Only recently, however, since the early 1990s, has its use been extended to QF observations of FCG under VA loading. This is mainly due to R&D by the USAF [74, 75] and DST [65, 66, 76–79], although others have observed exponential FCG behaviour under VA loading [80–82].

In particular, many DST investigations and studies have shown that approximately exponential FCG is a common occurrence for naturally-nucleating *lead cracks*, i.e.

the cracks leading to the first failures in airframe structures, components and specimens. The DST has published a detailed description of this phenomenon and the development of a Lead Crack Fatigue Lifing Framework (LCFLF) [65, 66]. The LCFLF has become an important *additional* tool, besides legacy standards [83, 84], to update the life assessments of RAAF aircraft [85].

The key elements of the LCFLF are (i) lead cracks nucleate from surface and near-surface material production discontinuities; (ii) these cracks begin to grow shortly after an aircraft enters service or soon after the start of testing; (iii) FCG is approximately exponential, see Eq. 1.14; (iv) significant portions of the FCG lives are spent within the short crack regime, in the crack depth range 0.01–1 mm; and (v) the small fraction of FCG life influenced by quasi-static fracture close to final failure is negligible. Full details of the LCFLF methodology and application are given in [65, 66]. Table 1.1 provides a stepwise general summary.

For actual service cracks the LCFLF needs:

(1) The sizes of the detected cracks and when they were detected.
(2) Estimates of the EPSs from which the cracks grew. If a crack origin is damaged a mean EPS for discontinuities typical of the material and surface finish is chosen.
(3) Estimates of a_{crit}.

With these inputs straight lines (extrapolations) are drawn from the EPS values to the a_{crit} values using log–linear coordinates: these represent the FCG curves. Finally, inspection intervals or safe lives are estimated. An example is given in Chap. 7.

In the foregoing sequence the usefulness of approximately exponential FCG derives from the near-linear a-N curves when plotted log-linearly. Two of the most noteworthy examples illustrating (a) the approximately linear behaviour and (b) straight line extrapolations to final detected crack sizes (*not* a_{crit}) will be given here:

(a) Figure 1.7 is a compilation of approximately exponential FCG curves obtained via QF for numerous locations in the AA2024-T851 and D6ac steel lower wing

Table 1.1 General summary of LCFLF methodology: after [65, 66]

• Assume immediate exponential FCG from material discontinuities characteristic of the production quality of the locations and areas to be assessed
• Convert discontinuity sizes (if available) into equivalent pre-crack sizes (EPS) or assume EPS values. The EPS metric is an essential part of the methodology, and also important in its own right [86, 87]
• Determine the final (critical) crack sizes, a_{crit}, characteristic of the locations and areas to be assessed
• Combine the previous steps to estimate the FCG lives to the critical crack sizes (i.e. log-linear extrapolation from the EPS to a_{crit}) and determine the lead crack FCG life in each location. **N.B**: this step must provide reliably conservative estimates of FCG lives
• Pool the lead crack lives for each location within each area, to calculate the average life of each area
• Apply appropriate scatter factors (SF) to the average life of each area to determine its safe crack growth life and/or inspection intervals

skin of an F-111 aircraft removed from service and tested under flight-by-flight block loading [88]. The FCG rates (gradients of the curves) were mostly similar, signifying that although the cracks grew at different locations, similar FCG rates pertained under similar loading conditions. This trend was also observed for full-scale tests on other structures [79].

Much more information is available [65, 66, 88] concerning the data in Fig. 1.7. Here we note two points of particular interest with respect to service failures:

- Under similar loading conditions the major source of scatter is the size of the initial discontinuities and hence the initial crack sizes.
- The approximate linearity of the FCG curves shows that it is reasonable to make linear interpolations of data sets when some QF measurements were not possible or available.

These points may be regarded as *additional* justifications for the following *earlier* example.

(b) Figure 1.8 shows four sets of FCG curves for several Aermacchi MB-326H AA7075-T6 wing spars [76, 89]. The data present a limited selection and interpolations of QF, from a total of 103 in-service fatigue cracks. The D17 curves are from the wing spars of a crashed aircraft (S/N A7-076): the upper curve is for the lead crack that caused failure of the left wing spar, resulting in loss of the wing. The D15, A14 and B19 curves are interpolations from the initial discontinuity sizes to the final crack sizes for other wings that had been removed from service. When processed via the steps indicated in Table 1.1, the results in Fig. 1.8 and numerous other results showed that continued operation of the fleet would require wing replacements [89].

N.B: The MB-326H crash occurred on 19th November 1990. This accident and the subsequent failure analysis may be regarded as a milestone in the evolution of aircraft structural integrity [90]. Specifically, this accident reinforced the recognition that QF is essential to (i) the structural integrity management of aircraft fleets [74, 90], and (ii) the analysis of full-scale, component and specimen fatigue tests [65, 66]. More details about the FCG analysis for this accident are given in Chap. 4.

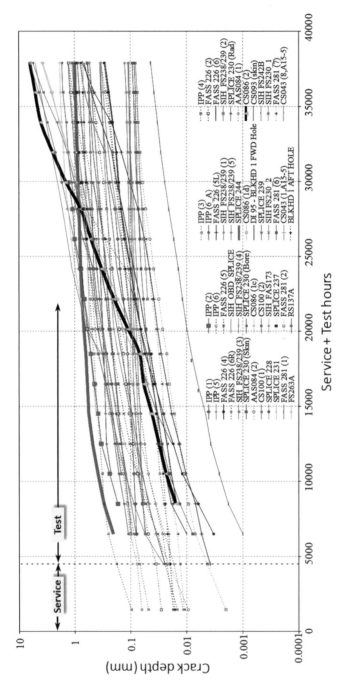

Fig. 1.7 Examples of QF-determined FCG curves from different locations in the lower wing skin of an F-111 test article removed from service [88]. CS = central spar; BLKHD = bulkhead; FASS = Forward Auxiliary Spar Station; IPP = Inner Pivoting Pylon; RS = Rear Spar; SIH = Sealant Injection Hole. **N.B**: The *final* lead crack, CS086 (2), did not lead until beyond 32,000 h

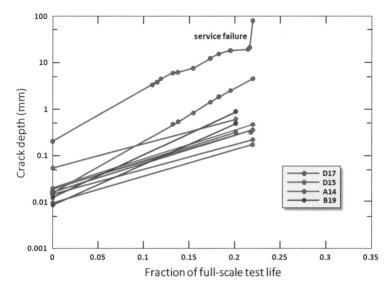

Fig. 1.8 Crack depths versus full-scale test life for fatigue cracks in the AA7075-T6 spars (D17 location) from MB-326H (S/N A7–076) and other locations in spars from three different aircraft: after [76, 89]

1.2.5.2 Exponential FCG: The Cubic Rule

The cubic rule represents a special category of lead crack FCG, using an extension of Eq. (1.14):

$$\textbf{FCG}: \quad a = a_0 \exp(\sigma_{REF}^{\alpha} \lambda N) \tag{1.17}$$

where α is a constant. From Eq. (1.17) the fatigue crack growth rate may be expressed as:

$$\textbf{FCGR}: \quad da/dN = a_0 \sigma_{REF}^{\alpha} \lambda \cdot \exp(\sigma_{REF}^{\alpha} \lambda N) = a \sigma_{REF}^{\alpha} \lambda \tag{1.18}$$

Equation (1.18) shows that the FCGR at any given crack length, a, is determined by $\sigma_{REF}^{\alpha} \lambda$. Then the FCGR ratio for two tests using the same load spectrum (same λ), but at two different reference stress levels may be expressed as:

$$(da/dN)_1 / (da/dN)_2 = (\sigma_{REF.1}/\sigma_{REF.2})^{\alpha} \tag{1.19}$$

Equation (1.19) implies that for a given load spectrum the lead crack growth rates obtained at one reference stress level may be used to predict the lead crack growth rates for a different reference stress level. This equation is very useful because tests with several load spectra and aerospace structural alloys, including aluminium alloys, a titanium alloy and a high strength steel, have shown that $\alpha \approx 3$ [91]. Owing to this result, Eq. (1.19) with $\alpha = 3$ has been designated *the cubic (stress-cubed) rule*.

Practical applications of the cubic rule include life predictions for structural repairs [92], and the use of QF-derived lead crack FCGR data from one location to predict the lead crack behaviour at other locations that experience the same load spectrum but at different stress levels. Both applications have increased the efficient use of QF data for life predictions. Two examples are given in Chap. 8.

1.2.5.3   Exponential FCGR

As stated at the beginning of Sect. 1.2.5, exponential FCGR equations have been found useful for *effectively* CA loading [71–73] as well as the VA block loading sequences in the EBA. The general exponential FCGR equation, Eq. (1.15), has been found useful for service and test fatigue failure analyses where the load histories were approximately CA. Examples are service loading of a helicopter rotor blade [73, 93]; full-scale test loading of a vertical stabilizer rib [94]; service and full-scale test loading of pressure cabin lap splices [71, 95]; and CA + marker load tests [72].

Depending on its range of crack size applicability, Eq. (1.15) may readily be integrated to obtain estimates of FCG lives:

$$N_f - N_i = (1/AB)(\exp(-Ba_i) - \exp(-Ba_f)) \tag{1.20}$$

where a_i and a_f are the initial and final specified crack sizes; and $N_f - N_i$ is the FCG life from a_i to a_f.

Figure 1.9 shows QF-derived FCGR data (fatigue striation spacings) for a Sikorsky S-61N AA6061-T6 rotor blade [93] and an exponential fit to the data. This example is chosen because the FCGR exponential fit was part of an FCG life analysis that also made use of LEFM da/dN versus ΔK data [73]. In other words, two FCG models were used in the analysis. However, these were not *alternatives*, but *complementary*, illustrating that the selection of FCG models should be based on 'whatever works best' in approximating reality, and that this can include a combination of models. More details about the FCG analysis are given in Chap. 9.

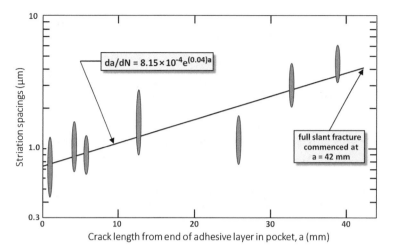

Fig. 1.9 Data envelopes for fatigue striation spacings, five measurements per position, for a fatigue cracked Sikorsky S-61N AA6061-T6 rotor blade that failed in flight. A best fit exponential FCGR equation is also shown. After [73, 93]

1.3 Summary

This Chapter has concisely reviewed several approaches to FCG analyses for metallic aircraft structures and components. Three are LEFM-based: LEFM cycle-by-cycle models, characteristic K approaches (including the EBA); and four are not: the LCFLF, the cubic rule, and general exponential equations for FCG and FCGR. All are *in principle* useful for FCG life predictions and service failure analyses, but there are some specific limitations, and also a basic assumption for the EBA and LCFLF:

(1) LEFM cycle-by-cycle models have been shown to be unreliable unless calibrated using realistic or actual service FCG data, especially QF data. In other words, so-called *blind* predictions, derived from *standard test data*, are unacceptable. Also, these models do not properly account for the FCG behaviour of short cracks.

(2) Characteristic K approaches (except the EBA) and general exponential FCG and FCGR equations have to be validated on a case-by-case basis. The EBA is an exception with respect to tactical aircraft VA load histories, owing to its derivation from extensive QF.

(3) The EBA and LCFLF have been extensively validated for tactical aircraft, (much) less so for other types of aircraft. However, the cubic rule, which is derived from the LCFLF, has proved useful in validating structural repairs for a maritime patrol aircraft [92].

(4) The EBA and LCFLF assume that FCG begins immediately an aircraft enters service. This assumption does not cover time-dependent crack nucleation events such as corrosion, fretting and accidental damage that can cause fatigue cracks

to nucleate and grow later in the service life. However, the EBA and LCFLF assumption is conservative, and in general well-justified for tactical aircraft.

As already mentioned, the selection of FCG models should be based on 'whatever works best' in approximating reality. This can include a combination of models, illustrated by use of (i) the cycle-by-cycle model AFGROW in combination with the EBA, and (ii) standard LEFM FCG data together with an exponential FCGR equation [73].

N.B: There is one very important point to be made in closing this Chapter. Concerns about environmental (corrosion) effects on fatigue cracking in aircraft structures have been expressed for decades, e.g. [96, 97], but none of the discussed FCG analysis methods explicitly consider environments other than ambient air. This might seem a serious omission, but although corrosion-enhanced FCG, i.e. corrosion fatigue, has been reported for aircraft service failures [98, 99], it appears to be unusual and not to be expected [95, 100–102]: see Sect. 2.3 in Chap. 2 also.

Chapter 2
Basic Information for Aircraft FCG Failure and Lifing Analyses

This Chapter presents some general and detailed information particularly relevant to FCG failure analyses. Three main topics are considered: (i) the importance of fatigue with respect to aircraft structural failures; (ii) fatigue nucleation and cracking; (iii) a summary of FCG lifing methods.

2.1 Importance of Fatigue for Aircraft Structural Failures

To begin, consider the information in Table 2.1 [103, 104]. This shows that fatigue failures predominate in metallic aircraft structures and are much more important than in general engineering. The relative importance of aircraft fatigue reflects the dynamic nature of the service loads and the generally high design stress levels required to achieve lightweight structures. Furthermore, fatigue failures are the main mechanical and structural contributors to serious aircraft accidents and incidents [103–107].

2.2 Fatigue Crack Nucleation, Discontinuities, Early FCG and Damage Tolerance

Two reviews of fatigue crack nucleation sites and the causes of fatigue-related *accidents* in metallic aircraft structures are the compilations by Campbell and Lahey [105] and Tiffany et al. [107], see Tables 2.2 and 2.3. These indicate that the main causes of airframe fatigue are stress concentrations and high local stresses. **N.B:** This conclusion may seem less than surprising, but it is most important to note from Table 2.2 the variety of stress concentrations implicated in fatigue-related accidents.

© The Author(s), under exclusive licence to Springer Nature B.V., part of Springer Nature 2019   21
R. Wanhill et al., *Fatigue Crack Growth Failure and Lifing Analyses*
for Metallic Aircraft Structures and Components, SpringerBriefs in Applied Sciences
and Technology, https://doi.org/10.1007/978-94-024-1675-6_2

Table 2.1 Failures in metallic aircraft structures and general engineering [103, 104]

Failure modes	Percentages of failures	
	Aircraft structures	Engineering industry
Corrosion	3–16	29
Fatigue	55–61	25
Brittle fracture	–	16
Overload	14–18	11
High temperature corrosion	2	7
Stress corrosion, corrosion fatigue, hydrogen embrittlement	7–8	6
Creep	1	3
Wear/abrasion/erosion	6–7	3

Table 2.2 Fatigue nucleation sites for aircraft accidents resulting from fatigue [105]

Nucleation sites	Stress concentrations	Numbers of accidents	
		Fixed wing	Rotary wing
Bolt, stud or screw	●	108	32
Fastener hole or other hole	●	72	12
Fillet, radius or sharp notch	●	57	22
Weld		53	3
Corrosion		43	19
Thread (other than bolt or stud)	●	32	4
Manufacturing defect or tool mark	●	27	9
Scratch, nick or dent	●	26	2
Fretting		13	10
Surface or subsurface flaw	●	6	3
Improper heat-treatment		4	2
Maintenance-induced crack	●	4	–
Work-hardened area		2	–
Wear		2	7

At this point it is necessary to broaden the scope to include fatigue cracking with *the potential* to cause accidents, and also more detailed information about fatigue nucleation sites and stress concentrations, since these are intrinsically related. This relationship is contemporarily implied and expressed by the term 'fatigue-nucleating discontinuities'.

Table 2.3 Fatigue causes for some aircraft accidents [107]

Fatigue causes	Numbers of accidents	
	Airframes	Engine discs
Unanticipated high local stresses (possibly combined with final manufacturing defects)	11	–
Manufacturing defect or tool mark	3	2
Material defect	2	1
Maintenance deficiencies	6	–
Abnormally high fan speed		1

2.2.1   Fatigue-Nucleating Discontinuities in Aircraft Structures

There are two categories of fatigue-nucleating discontinuities in aircraft structures: (i) discontinuities pre-existing before the aircraft enter service; and (ii) damage that occurs during service. *Typical* examples of pre-existing discontinuities are listed in Table 2.4. Most are for aluminium alloys, reflecting their widespread use in metallic aircraft structures. All of these discontinuities are capable of nucleating fatigue cracks very early in the service life, especially in the presence of high local stresses and stress concentrations, see Tables 2.2 and 2.3. A detailed discussion, with fractographic illustrations, of this category of discontinuities is given in [65,108]. Some examples are also given in Chap. 3, Sect. 3.4.

The second category of fatigue-nucleating discontinuities includes time-dependent service damage, notably corrosion and fretting, and incidental damage, e.g. during maintenance, see Tables 2.2 and 2.3. Of these, fretting is known to cause cracking very early in the fatigue life, e.g. [109–111].

N.B: Although manufacturing and material defects are listed in Tables 2.2 and 2.3, they are not included in Table 2.4 owing to their anomalous nature and relative rarity [112]. However, the irony is that a material defect was responsible for '*the most infamous crack in aviation history*' [107]. This refers to the 1969 crash of a General Dynamics F-111A. As mentioned in Sect. 1.1, this crash was partly responsible for the USAF introducing its Damage Tolerance philosophy [23, 24]: see also [90].

There are several important characteristics associated with typical pre-existing fatigue-nucleating discontinuities:

(1)   Their sizes are often much less than 1 mm, resulting in FCG from crack depths in the range 0.01–0.1 mm. Some examples are given in [108].
(2)   As mentioned in Sect. 1.2.5.1, a significant amount of FCG occurs within the short crack regime. For aircraft structural alloys this regime may be considered to extend to crack depths of about 0.5–1 mm. Precise values cannot be given, since short crack behaviour is complex [113], depending on numerous factors including the crack geometry, material, surface condition, local stress concentration and overall stress level.

Table 2.4 Typical examples of pre-existing fatigue-nucleating discontinuities [108]

Discontinuity sources	Specifics	Aircraft type	Fatigue conditions
Poorly finished holes	Poor drilling	Aermacchi MB326	FSFT*
	Poor drilling	Airbus A380 MLB**	FSFT
	Scoring from fastener	Aermacchi MB326	Service
	Poor de-burring	Aermacchi MB326	Service
	Machining tears/nicks	Aermacchi MB326	Service and FSFT
	Machining tears/nicks	Dassault Mirage III	Service
	Machining tears/nicks	Boeing F/A-18A/B	Coupon test
	Machining tears/nicks	Lockheed P3C	FSFT
Surface treatments	Etch pits	Boeing F/A-18A/B	Coupon tests
	Etch pits + machining	Boeing F/A-18A/B	Service
	Intergranular "penetration"	Boeing B747	Service
	Chemical milling + peening	G.D.*** F-111	Service and FSFT
	Pickling	Lockheed P3C	FSFT
	Peening laps and cuts	Boeing F/A-18A/B	Coupon tests
Porosity	Thick plate porosity	Boeing F/A-18A/B	FSFT
Constituent particles	Cracked particles	–	Coupon test
	Varying shapes	Boeing F/A-18A/B	Coupon tests
	Particles + poor machining	Dassault Mirage III	FSFT

* FSFT = Full-Scale Fatigue Test ** MLB = MegaLiner Barrel
***G.D. = General Dynamics (now Lockheed Martin)

(3) FCG begins almost immediately after an aircraft enters service, essentially immediately for lead cracks, and the crack growth behaviour is approximately exponential. These characteristics are also mentioned in Sect. 1.2.5.1.

All three of the above points have important implications for the estimation and prediction of FCG lives, as discussed in Sect. 2.2.2.

2.2.2 Early FCG and Damage Tolerance

The USAF Damage Tolerance guidelines were introduced in the mid-1970s [23, 24] with the objective of estimating FCG from pre-existing damage (cracks or crack-like flaws). As mentioned in Sect. 1.1, the guidelines covered the issues of *safety* and *durability*, for which different initial crack/flaw sizes had to be assumed. Safe FCG lives were to be estimated with initial crack/flaw sizes ranged from about 0.5–6 mm [23]. This size range was, and is, large enough to use LEFM-based macrocrack growth data and models for the estimates.

However, *durability* was problematical since it required estimates of fatigue-determined economic lives from much smaller initial crack/flaw sizes. In the 1970s there was very little information about the sizes of typical fatigue-nucleating discontinuities and early FCG behaviour from these discontinuities. Instead recourse was necessarily made to assuming an initial crack/flaw size or estimating equivalent initial flaw sizes (EIFS) by back-extrapolation from macrocrack test data, using LEFM macrocrack growth models for the back-extrapolation and also for subsequent durability FCG life estimates. In both cases there was an implicit assumption that macrocrack growth models would give reasonably reliable results. Later it was recognised that back-extrapolation of macrocrack test data is unreliable for the short crack regime, e.g. [35, 37, 114], rendering the derived EIFS values and FCG lives highly questionable. This problem increased in scope when it was also recognised that *safety* and *durability* are linked for ageing aircraft.

The uncertainties surrounding the original EIFS concept led the DST to characterise actual fatigue-nucleating discontinuities using QF, particularly during development of the lead crack approach (LCFLF) discussed in Chap. 1. As a result, the DST introduced the equivalent pre-crack size (EPS) concept. This concept avoids (i) extensive back-extrapolation of macrocrack test data and (ii) use of macrocrack growth models. Instead the EPS uses QF-validated exponential FCG relationships covering the short crack regime: see Chap. 3, Sect. 3.4.1, for more information on the EPS.

N.B: It was always the USAF's intention to use the actual sizes of fatigue-nucleating discontinuities, but—as noted above—such data were not available in the 1970s. Recently, however, a USAF—DST extensive programme of VA spectrum testing has shown that the EIFS and EPS concepts can be reconciled, i.e. they become essentially equivalent, when using exponential or LEFM-based FCG models that have been QF-validated for the short crack regime [87]. This result reinforces the point made in Sect. 1.3 that the selection of FCG models should be based on 'whatever works best' in approximating reality.

An additional message from the above-mentioned USAF and DST programme is that international collaboration is important. In this respect the HOLSIP programme [113] is especially noteworthy. As befits its name, HOLSIP (HOListic Structural Integrity Process) is a very broad programme first proposed in 2001. Since then it has been under development with participants from Canada, USA, UK, Australia, Japan, Poland and the Netherlands. The FCG aspects of HOLSIP are summarised in Sect. 2.3.

2.3 Summary of FCG Lifing Methods

A summary of FCG lifing methods for metallic airframe structures is given in Table 2.5: see p. 27. The USAF and DST approaches have already been discussed. HOLSIP is included in this summary for completeness. Some additional information about these methods is given here:

(1) USAF Damage Tolerance: This method has been used for many USAF air-craft since the mid-1970s, and has been effective in ensuring structural *safety*, whereby the *assumed* initial crack/flaw sizes were large enough for estimations of FCG using LEFM models based on macrocrack growth behaviour. However, this method, as originally implemented, has been less satisfactory for analyses of structural durability, i.e. estimating economic lives based on FCG from (much) smaller initial crack/flaw sizes. There were two main problems:

  • In the 1970s there was little information about the sizes of small fatigue-initiating discontinuities and early FCG from them. Hence it was necessary to estimate EIFS values and assume that these would provide reliable starting points for FCG.
  • The EIFS values were estimated by back-extrapolation from macrocrack FCG data, using LEFM macrocrack growth models. Later on, following research in the 1970s and 1980s [26–32], it became clear that the EIFS values were unreliable because they were well within the *short crack regime*, and the FCG behaviour in this regime cannot be reliably estimated using LEFM models based on macrocrack growth behaviour. This also meant that estimates of the *durability* FCG lives were unreliable [35].

In response to this situation the USAF recently participated in a collaborative programme of VA spectrum testing with the DST [87]. This programme showed that realistic EIFS values can be obtained via FCG models applicable in the *short crack regime*.

(2) DST approach: Advances in QF since 1990 provided many data for determining the sizes of typical fatigue-nucleating discontinuities and early FCG from them, particularly for tactical aircraft. Both exponential and LEFM-based FCG models have been developed to use the discontinuity data, converted into EPS values, for FCG life estimates. An important aspect of these estimates is that so-called *lead cracks* often exhibit exponential FCG, and this has led to development of the LCFLF [65,66].
N.B: The above-mentioned collaborative programme with the USAF showed that the EPS and EIFS concepts are essentially the same when based on FCG models that are QF-validated in the *short crack regime* [87].

(3) HOLSIP approach: This is a broad and ambitious international programme that includes more than FCG analyses. Table 2.5 shows the main elements and objectives of the HOLSIP FCG approach. These are similar to those of the USAF and DST approaches, but there are two important differences:

- Firstly, environmental effects are considered. These include corrosion-*induced* FCG and corrosion-*enhanced* FCG, i.e. corrosion fatigue. Both may be appropriate to a holistic approach, which considers design and maintenance issues as well as failure analyses. However, as remarked at the end of Sect. 1.3 in Chap. 1, although corrosion-*enhanced* FCG has been reported for aircraft service failures [98, 99], it appears to be unusual and not to be expected [95, 100–102].
- Secondly, HOLSIP defines discontinuity states (DS) that 'evolve' from initial discontinuity states (IDS). The IDS concept is essentially the same as the EPS and updated EIFS [87] concepts, and may therefore be used for FCG estimations from the beginning of service use or after time-dependent fatigue nucleation has occurred.

The DS represent increased size distributions and average discontinuity sizes owing to fatigue and/or corrosion during service. This IDS → DS concept is intended to provide DS starting conditions for FCG estimations at some stage, or stages, in the service life: for example depot-level maintenance with non-destructive inspection (NDI) that quantifies and puts upper limits on the current DS. This time-dependent aspect of FCG is important, but not a *primary* consideration for the present book. See Chap. 10 for a simple (but exceptional) practical example of avoiding the influence of corrosion on fatigue crack nucleation during service.

Table 2.5 Summary of FCG lifing methods for metallic airframe structures

USAF original Damage Tolerance approach [23,24]
• Equivalent initial flaw sizes (EIFS) derived from back-extrapolation of macrocrack test data
• LEFM long crack growth models (non-interaction, yield zone, crack opening, strip yield) to derive variable amplitude (VA) crack growth from constant amplitude (CA) data
• Possible use of crack opening models for short cracks, e.g. FASTRAN [53]
• Mainly deterministic: stochastic approach also possible
DST approach and LCFLF [65,66]: implemented by the RAAF
• Actual initial discontinuity/flaw sizes and their equivalent pre-crack sizes (EPS)   • Actual short-to-long crack growth data using Quantitative Fractography (QF)   • Data compilations to establish FCG empirical relationships: typically exponential
• Deterministic ("upper bound") estimates of FCG, including *lead crack* growth
• Scatter factors
HOLSIP approach [115]
• Initial discontinuity sizes (IDS) that 'evolve' to discontinuity sizes (DS) during service
• Evaluation and selection of marker load strategies for QF of short-to-long crack growth
• Actual short-to-long long crack growth using marker loads and QF
• Establishment, validation and choice of appropriate crack growth models and "laws"
• Deterministic ("upper bound") and stochastic estimations of FCG
• Environmental effects, notably corrosion

Chapter 3
Quantitative Fractography (QF) for FCG Analyses

3.1 Introduction

The study of fatigue fractures has a long history [116]. Macroscopic fatigue progression markings on fracture surfaces were mentioned as such by 1926 [117]. Also, as mentioned at the beginning of Chap. 1, FCG analyses began in the 1950s [1–6], notably with the study of microscopic fatigue striations and recognition that they generally represent cycle-by-cycle progression of a crack [7, 8]. There is an extensive literature on fatigue striations, much of which concerns aluminium alloys and/or the mechanism(s) of striation formation.

In the following two Sections we first discuss FCG analyses based on striation measurements before considering progression markings. This is because striation-based QF FCG analyses have significant limitations, whereas similar analyses using progression markings are generally much more useful. Section 3.4 discusses typical fatigue-nucleating discontinuities and the EPS concept, and Sect. 3.5 briefly surveys QF techniques.

3.2 Fatigue Striation FCG Analyses

Fatigue striations are commonly confined to Region II of the da/dN versus ΔK diagram, see Fig. 1.1 in Chap. 1. However, this does not mean that they always occur in Region II or are clearly observable. Aerospace materials for which striations are readily visible include aluminium alloys, stainless steels and nickel-base superalloys, see Fig. 3.1a–c. On the other hand, for high strength steels and titanium alloys Fig. 3.1d–g show that the alloy microstructure can have a strong influence on the FCG topography, such that striations are not observable or else found only in localised areas on the fracture surfaces.

© The Author(s), under exclusive licence to Springer Nature B.V., part of Springer Nature 2019
R. Wanhill et al., *Fatigue Crack Growth Failure and Lifing Analyses*
for Metallic Aircraft Structures and Components, SpringerBriefs in Applied Sciences
and Technology, https://doi.org/10.1007/978-94-024-1675-6_3

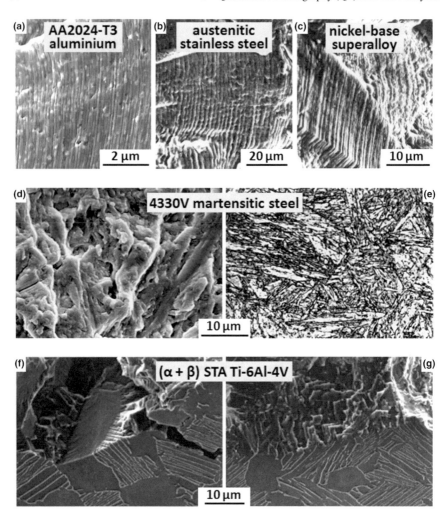

Fig. 3.1 Region II FCG: **a–c** fatigue striations; **d** fractograph of microserrated acicular ridges and **e** underlying microstructure; **f**, **g** precision sections showing (**f**) cleavage-like fracture through an α grain and (**g**) blocky fracture through a transformed β grain. *Source* Barter SA, Lynch SP, Wanhill RJH (2016) Failure Analysis of Metallic Materials: A Short Course, Netherlands Aerospace Centre NLR, Amsterdam, the Netherlands, and Defence Science and Technology Group DSTG, Melbourne, Australia. Available via ResearchGate: www.researchgate.net

Besides microstructure-dependent observability, there are other caveats:

(1) Striation spacings do not necessarily correspond to the macroscopic crack growth rates, even under CA loading. At high growth rates the striation spacings underestimate the macroscopic crack growth rates because microvoid coalescence owing to tensile tearing also contributes to crack growth. On the other hand, at low growth rates there appears to be a lower limit to striation spacings that leads to overestimating the macroscopic crack growth rates [116]. In part this lower limit may be due to difficulties in detecting and resolving striations on some fracture surfaces [118].

A most important consequence of a lower limit to reliable measurements and detecting and resolving striations is that it is usually not possible to trace them back to typical fatigue-nucleating discontinuities. This means that reliable early FCG data, which represent significant amounts of the FCG lives, often cannot be obtained from striation spacings.

(2) Realistic VA loading sequences usually make striation spacing measurements unusable, since they are greatly affected by the differing load excursions and any variations in mean loads. QF measurements must instead rely on progression markings, whose 'readability' in terms of their relationships to the VA loading sequences may be difficult to establish, see Sect. 3.3.

(3) The environment can have an important influence on the visibility of striations, which are best visible from fatigue in mild environments like ambient air. More aggressive environments such as salt water or high temperature air can damage and eventually obscure the original fracture surfaces by corrosion and oxidation. On the other hand, changes in the environment can cause changes in crack growth behaviour that result in progression markings. For example, service fatigue cracks sometimes begin subsurface, typically owing to shot peening the surface. However, when a crack breaks through to the surface the transition from fatigue *in vacuo* to environmental fatigue can produce a well-defined progression marking, even when striations are not readily visible after the breakthrough [119].

Notwithstanding these issues, if the service (or test) fatigue loading is essentially CA, i.e. without progression markings, then striation spacing measurements may be the only possible way of using QF to obtain FCG data for failure analyses. Two examples are given in Chap. 9 (Fokker 100 full-scale fatigue test) and Chap. 10 (Sikorsky S-61N rotor blade service failure).

3.3 Progression Marking FCG Analyses

There are basically two categories of progression markings, natural and artificial:

(1) Natural progression markings may have various causes, the most common being an abrupt or sustained significant change in cyclic loading. Other causes include

start–stop cycles in aeroengines, overloads, tensile crack jumping due to peak loads [120, 121], environmental changes, and transitions in fatigue crack topography owing to changes in cyclic plasticity, notably in titanium alloys [119, 122]. This category of progression markings may or may not be useful for QF, depending on whether they can be correlated to (a) specific recorded events during FCG, e.g. service load recordings; or (b) somewhat exceptionally to well-defined fracture mechanism changes and topography transitions that can be quantitatively related to the crack tip stress intensity factors [119, 121]. Service failures illustrating (a) and (b) are given in Chap. 4 (Aermacchi MB-326H wing spar) and Chap. 11 (Westland Lynx rotor hub), respectively.

(2) Artificial progression markings are usually single or block marker loads inserted into CA and VA loading sequences. A further development is to derive a realistic VA loading sequence consisting entirely of simulated flights that result in QF 'readable' progression markings. Examples of all these types have been extensively reviewed [72]. This review also provides guidelines for obtaining good QF 'readability' while not significantly affecting the overall FCG process. Artificial progression markings can be most useful for full-scale, component and specimen fatigue tests. In fact, they have been essential in the development of the EBA, LCFLF, and the cubic rule (Chap. 1), and the EPS concept (Chap. 2).

Figure 3.2 shows two very different examples of the QF capabilities with respect to VA spectrum tests: (i) the Mirage wing spectrum contained several features resulting in progression markings, including high loads in each block of 200 simulated flights, different types of flights, and ground-air-ground cycles for each flight; (ii) the F/A-18 aft fuselage spectrum contained numerous high loads that resulted firstly in progression markings and later in progression markings and tensile crack jumping (the grey tongue-shaped areas).

A schematic illustration of QF measurements from progression markings is given in Fig. 3.3. This schematic is idealised, since actual fracture surfaces may have local areas where the topography is rougher and progression markings are less visible. Even so, the crack fronts represented by progression markings may be assumed to have a regular geometry, such that it is acceptable to interpolate across any non-observable segments. Besides crack depth measurements the crack widths and the shapes and sizes of the fatigue-nucleating discontinuities should be determined, see Sect. 3.4. These basic measurements may then be used to assist in obtaining FCG plots, particularly log–linear plots of crack depth versus number of simulated flights or flight blocks to check— in the light of discussions in Chap. 1—whether FCG might be approximately exponential.

Chapters 6–8 discuss some test results, obtained mostly from QF of artificial progression markings, important to the development of the EBA, LCFLF, and cubic rule concepts.

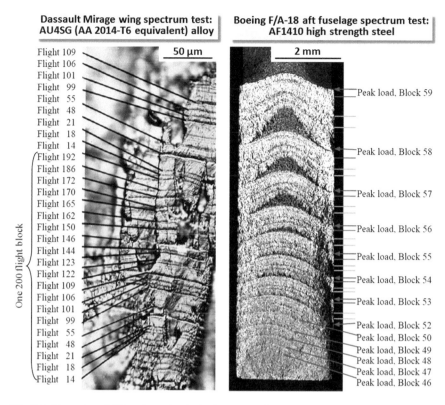

Fig. 3.2 Examples of QF analyses (optical microscopy) for VA spectrum tests. The Mirage fractograph is from a full-scale test [123]; the F/A-18 fractograph is from a specimen test [124]

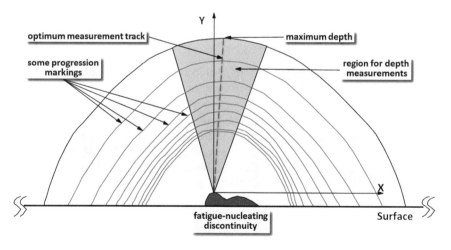

Fig. 3.3 Schematic illustration of QF measurements from progression markings

3.4   Fatigue-Nucleating Discontinuities and Equivalent Pre-crack Sizes (EPS)

The reader is referred to Chap. 2, Sect. 2.2, for information about the types of typical fatigue-nucleating discontinuities, see Table 2.4. These are usually very small, such that FCG begins at crack depths in the range 0.01–0.1 mm. Furthermore, FCG can begin very early—essentially at the beginning—of service and therefore account for a significant amount of the fatigue life. Hence it is important to quantify the discontinuity sizes and obtain insights into the commencement of FCG from them. A few examples are illustrated in this Section before discussing methods of characterising discontinuities for FCG analyses.

Figure 3.4 compares optical 'deep focus' and scanning electron microscopy (SEM) fractographs of three types of fatigue-nucleating discontinuities in AA7050-T7451 aluminium alloy plate components from RAAF-operated Boeing F/A-18 aircraft. The deep focus fractographs were obtained via digital processing of stacks of optical fractography images [125]. All of the images represent in-service FCG followed by limited CA testing and then VA block loading testing that resulted in well-defined progression markings. There are three main points to note:

(1) The discontinuity sizes in the depth direction range from 0.03 mm for the etch pit to 0.07 mm for the inclusion (preferably described as a constituent particle, since such inclusions are inherent to commercial high-strength aluminium alloys). **N.B**: Parts of the inclusion have broken away during the cracking process, resulting in voids.
(2) The progression markings are more readily observable from the deep focus images. This is especially the case for Fig. 3.4c, where the fatigue fracture surface has been partially obscured by corrosion.
(3) Progression markings show that semi-elliptical crack fronts started to form already at crack depths less than 0.02–0.03 mm, i.e. depths less than those of the discontinuities.

Figure 3.5 shows *two* examples of information about early FCG in carefully prepared specimens of AA2024-T3 aluminium alloy subjected to VA flight simulation loading. Fatigue cracks nucleated at surface-breaking constituent particles, typically at crack depths less than 0.01 mm. Again, there was the tendency to form semi-elliptical crack fronts, but the diagram in Fig. 3.5 shows that the crack front shapes were initially shallow and gradually progressed to an equilibrium shape predicted by stress intensity factor analysis. This equilibrium shape was achieved at a crack depth of about 0.05 mm. **N.B**: (i) the arrow in the fractograph points to a void, about 0.006 mm (6 μm) deep, where the constituent particle originally was; (ii) progression markings are not observable on the left-hand side of the fractograph owing to a very rough fracture topography (reason or reasons unknown).

From the foregoing examples and discussion about them it is already evident that defining an initial crack size at which FCG may be assumed to commence is not an easy task. Other examples, also for aluminium alloys, demonstrate how complex the

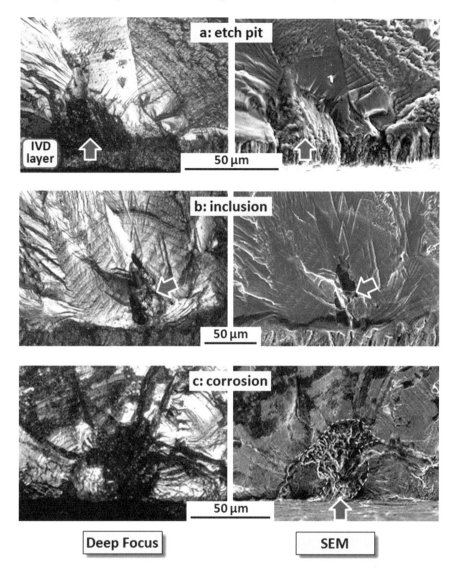

Fig. 3.4 Examples of deep focus (left) and SEM (right) fractographs of fatigue-nucleating discontinuities and early FCG in AA7050-T7451 plate. The etch pit was due to chemical etching before applying an ion vapour deposition (IVD) aluminium coating for corrosion protection. This coating filled the etch pit. *Source* Barter SA, Defence Science and Technology Group DSTG, Melbourne, Australia

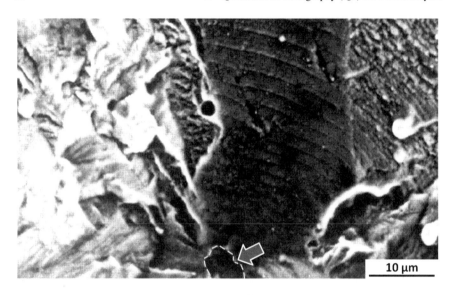

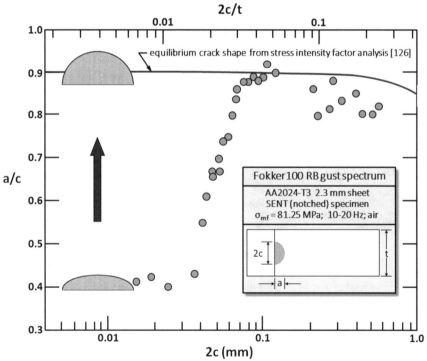

Fig. 3.5 Examples of FCG from surface-breaking constituent particles (inclusions) in specimens of AA2024-T3 sheet tested under Fokker 100 transport aircraft flight simulation loading: after [127]

shapes of fatigue-nucleating discontinuities can be, particularly for near-surface and surface-breaking porosity [108], constituent particles [108, 128, 129] and corrosion pits [130].

3.4.1 Defining the Initial Crack Sizes for FCG Analyses

The possibilities for defining initial crack sizes for FCG analyses are:

(1) Conversion of the fatigue-nucleating discontinuity size into a $\sqrt{area}$ crack size. For 3D cracks the $\sqrt{area}$ crack size correlates with the maximum stress intensity factor along the crack front [131]; and for a semi-elliptical surface crack with a/c < 1.0, as is usually the case, this corresponds to the crack depth [132]. **N.B**: The $\sqrt{area}$ conversion is an approximation that appears to be widely applicable for many types of discontinuities in different classes of alloys, including aluminium alloys and steels [131].

(2) Back-extrapolation of FCG curves determined from progression markings to the beginning of crack growth. This does not require the size of the fatigue-nucleating discontinuity, and can be used, *as a necessary resort*, if the fatigue origin is damaged. This method is more likely to be reasonably accurate if early FCG is approximately exponential. Chapter 5 gives an example for a steel bearing cage from a turboprop engine.

(3) A judicious combination of (1) and (2), i.e. when the fatigue origin is *not* damaged and the fatigue-nucleating discontinuity size can be measured.

(4) The EPS approach. This is an important sophisticated version of (3), that has involved the collation and statistical analysis of more than 900 FCG curves with respect to several types of fatigue-nucleating discontinuities [86]. The EPS is essential to the LCFLF, as mentioned previously, and is the most well-established method of estimating the initial crack sizes for FCG analysis. At present the EPS analysis is restricted to one alloy, AA7050-T7451, and has been used in FCG analyses and service life assessment for the fleet of Boeing F/A-18 aircraft operated by the RAAF. See Chap. 7 for a detailed example.

Table 3.1 compares the derived EPS depths and measured discontinuity depths at the 50th percentile (median) level, since the median values of EPS have been used for the F/A-18 aircraft LCFLF analyses. The median depths agree reasonably well except for constituent particles and porosity: the much smaller EPS depths mean that constituent particles and porosity behave in a much less crack-like manner when acting as fatigue nucleation sites. This is also the case for corrosion pits [130].

(5) A combination of (2) and (4) as an improvement on (3), but in this case restricted to high-strength aluminium alloys. This method assumes that (i) the fatigue-nucleating discontinuity in a high-strength aluminium alloy belongs to one of the categories listed in Table 3.1; (ii) the appropriate EPS/discontinuity size ratio in Table 3.1 is transferable from AA7050-T7451 plate to this other alloy; and (iii) the discontinuity depth can be measured so that its EPS can be obtained from this EPS/discontinuity size ratio.

Table 3.1 EPS and discontinuity depths (mm) and ratios for AA7050-T7451 aluminium alloy plate at the 50th percentile (median) level [86]

Discontinuity types	EPS	Discontinuity	Ratio: EPS/Discontinuity
OEM[a] etch pits	0.016	0.015	1.07
DST etch pits	0.009	0.011	0.82
Constituent particles	0.011	0.024	0.46
Mechanical damage	0.033	0.030	1.10
Peening damage	0.025	0.022	1.14
Porosity	0.007	0.049	0.14

[a]OEM = Original Equipment Manufacturer (Boeing)

3.5 QF Techniques for Fatigue Fracture Surfaces

The techniques that are, or have been, used for QF of fatigue fracture surfaces are listed here with their (very) approximate magnification ranges:

- Optical microscopy, including deep focus [125] 1–1500×

- Scanning Electron Microscopy (SEM) 10–20,000×
  - field emission gun (FE-SEM) 10–40,000×
  - secondary electron (SE) mode
  - back-scattered electron (BSE) mode.

- Transmission Electron Microscopy (TEM) of replicas 1000–50,000×
  - very rarely used nowadays.

Examples of the required optical and scanning electron microscopy (FE-SEM) equipment are shown in Fig. 3.6. The deep focus microscope has a very rigid construction; long working distance objectives up to 150×; a low noise digital camera; a computer-controllable stepping motor driven stage (X, Y, Z directions in steps < 1 μm); and a modified version of the open source ImageJ program to drive the stage and camera.

The FE-SEM has better resolution than a standard SEM (not always necessary in the present context); low keV operation, which is advantageous when insulating materials like surface coatings are associated with fatigue fractures; and—like a standard SEM—a range of available analysis equipment to choose from, e.g. for identifying constituent particles at fatigue origins.

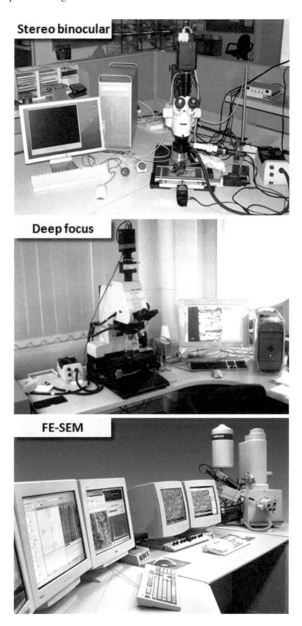

Fig. 3.6 Examples of stereo and deep focus optical microscopy equipment and an FE-SEM. *Source* Barter SA, Defence Science and Technology Group DSTG, Melbourne, Australia

3.5.1 Optical Fractography

Conventional and deep focus optical fractography have the advantages, compared to SEM, that colours, slight tilts and undulations are readily detected, and that contrast can be enhanced by oblique illumination and interference contrast. The general problem of limited depth of field is compensated by the deep focus technique which, as mentioned with respect to Fig. 3.4, involves digital processing of stacks of optical fractography images [125].

At low magnifications (10–200×) stereo binoculars with camera-recording facilities may be used to obtain an overall impression of a fatigue fracture surface, including any prominent progression markings. Sometimes it may be possible to track the entire FCG history from low magnification optical fractographs, e.g. the F/A-18 steel fractograph in Fig. 3.2. However, it is more often useful to prepare low magnification fracture surface photomontages as guides for examination at higher magnifications using deep focus or SEM fractography. A similar suggestion, with respect to SEM fractographs, is made in Sect. 3.5.2.

3.5.2 Electron Fractography and Optical Fractography

The main advantages of using SEM and FE-SEM fractography are the much greater depth of field and better resolution, particularly if it is necessary or useful to measure fatigue striation spacings. However, magnifications above about 1500× may or need not be required when tracking progression markings, and these are usually more visible from deep focus fractography, as shown already by Fig. 3.4. This difference illustrates a disadvantage of electron fractography, namely the general lack of contrast in the standard SE mode and a 'flattening' of the fracture surface appearance. This latter aspect can be remedied by taking stereopairs of fractographs.

Whether to use deep focus or SEM fractography depends on the availability of a deep focus microscope, since these are less commonly part of the microscopy inventory; and also on personal choice. There is something to be said for a combination of their advantages, e.g. obtaining a *relatively* low magnification SEM preview of a fracture surface, to be used as a guide, before continuing at higher magnifications with deep focus or SEM, or both.

3.6 A Cautionary Note

Interpretation and quantification of fatigue fracture surfaces requires much expertise and experience, even when facilitated by materials that readily show fatigue striations and progression markings. For example, see the Mirage aluminium alloy fractograph in Fig. 3.2. A *sine qua non* is that failure analysis investigators should not delegate detailed fractographic examinations until having viewed the fracture surfaces themselves and specifying what should be done. This can take considerable time, sometimes days rather than hours.

Chapter 4
Aermacchi MB-326H Wing Spar (1990): Exponential FCG Analysis

4.1 Introduction

In November 1990 an RAAF Aermacchi MB-326H jet trainer, tail number A7-076, crashed into the sea off the east coast of Australia. The crash occurred owing to loss of the left wing during a 6.5 g manoeuvre. An extensive search resulted in recovery of the right wing, centre section and stub of the left wing from the ocean [76]. Figure 4.1 shows the origin of the left wing failure, which began from a poorly-drilled bolt hole in the flange of the left wing lower spar cap (AA7075-T6 aluminium alloy). This hole should have been drilled to a flat bottom and then reamed, but the drill had penetrated through to the inside surface of the spar flange, resulting in severe stress concentrations at the bottom of the hole [89].

The spar was part of a replacement programme that included the wing centre sections and spars of the MB-326H fleet. This programme, called LOTEX (Life-Of-Type-EXtension), was done in the mid-1980s by the Commonwealth Aircraft Corporation. Since replacement the aircraft had accumulated 1904 flights, representing 2188 airframe flight hours. This was only 70% of the estimated fatigue life based on the full-scale fatigue test done in 1974/5. **N.B**: This fatigue life was obtained via the Safe Life philosophy, i.e. no measurable fatigue cracking during the service life. This approach has been superseded by the Damage Tolerance philosophy, see point (2) in Sect. 1.1 of Chap. 1.

4.2 FCG Analysis for the Failed Spar

The A7-076 failure analysis may be regarded as a major precursor to developing the LCFLF. This is in addition to the accident and failure analysis representing a milestone in the evolution of aircraft structural integrity [90], see Sect. 4.4 and the note at the end of Sect. 1.2.5.1 in Chap. 1.

R. Wanhill et al., *Fatigue Crack Growth Failure and Lifing Analyses for Metallic Aircraft Structures and Components*, SpringerBriefs in Applied Sciences and Technology, https://doi.org/10.1007/978-94-024-1675-6_4

43

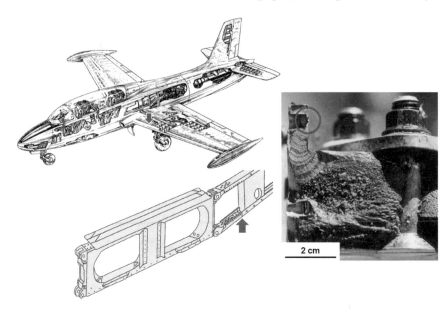

Fig. 4.1 Origin of the MB-326H (A7–076) left wing failure, a poorly drilled bolt hole (circled) in the lower spar cap flange. FCG progression markings are evident on the spar cap fracture surface

Optical fractography was carried out on the fatigue crack in the failed left wing lower spar cap and, after NDI, the opened-up longest fatigue crack in the right wing lower spar cap. The fatigue cracks showed progression markings owing to peak loads, and these markings later changed into tensile crack jumping. After considerable trial and error [76], a good match was found between progression markings and +6g peak loads registered by 'g' meters installed on the aircraft. Figure 4.2 shows the match for the right wing spar cap. The tensile crack jumps, which were due to peak loads, aided the fractographic analysis. The FCG curves for both cracks were found to be approximately exponential, and this led to extrapolating them back to the initial defect sizes, see Fig. 4.3. The reasonable fit suggests that FCG began soon after the aircraft re-entered service after the LOTEX programme. **N.B**: Use of initial defect sizes has been superseded by the EPS in later work, see Sect. 3.4.1 in Chap. 3.

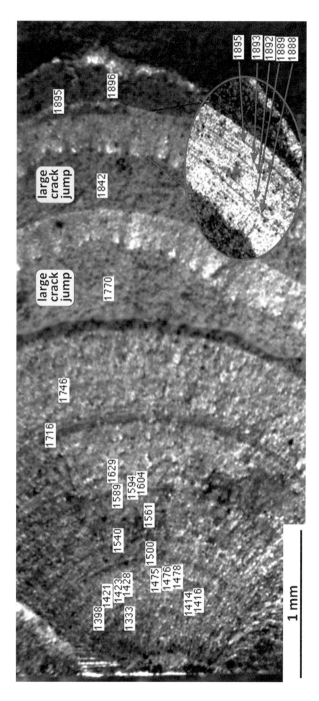

Fig. 4.2 Optical fractograph of the longest fatigue crack in the right wing lower spar cap from MB-326H (S/N A7-076). The numbers represent service flights matched to +6g peak loads, with a special 'window' near the end of crack growth. Two large tensile crack jumps are also indicated. *Source* Goldsmith NT, Defence Science and Technology Group DSTG, Melbourne, Australia

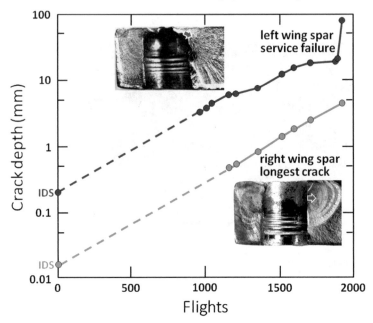

Fig. 4.3 Crack depths versus service flights for fatigue cracks in the AA7075-T6 lower spar caps from Macchi MB-326H (S/N A7-076): IDS = initial defect size; arrows point in the crack growth directions. After [76,89,133]

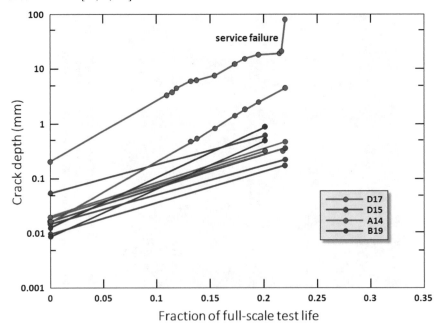

Fig. 4.4 Crack depths versus full-scale test life for fatigue cracks in the AA7075-T6 spars (D17 location) from MB-326H (S/N A7-076) and other locations in spars from three different aircraft. After [76,89,133]

4.3 Further Examination of Wing Spars

Detailed investigation of the A7-076 failed spar cap revealed manufacturing discon-tinuities in many holes, making it probable that others would be present in spars from other aircraft in the fleet. An NDI procedure was developed and employed for fleet-wide detection (within the NDI limits) of any such cracks. This resulted in crack indications in three other wings besides those from the A7-076 accident. In addition, detailed (teardown) examination of the cracked lower spar cap from the right wing of A7-076 showed that there were many cracks growing from *normal quality* structural details [89]. This meant that the Safe Life of the entire fleet needed to be reassessed, and the results led to a fleet recovery programme. A full description of these activities is provided by Clark et al. [89] and summarised in Sects. 4.3.1 and 4.3.2.

4.3.1 Safe Life Reassessment

The Safe Life reassessment was based on teardown, NDI and QF of detected fatigue cracks for five high-life wings besides the A7-076 left wing stub and right wing. Figure 4.4 shows a selection of the FCG plots for some of the 'worst case' cracks compared with those from the A7-076 left wing stub and right wing. The assumption made for these plots was immediate in-service exponential FCG from the measured initial discontinuities. Also, the FCG lives were converted into Safe Life damage units used for the full-scale test life. This accounted for different aircraft having different service load histories.

FCG plots for the 'worst case' cracks were extrapolated to predicted failure points (critical crack lengths) according to a residual strength criterion of 6.5g [89]. The resulting FCG life predictions were (i) converted into Safe Life damage units, (ii) pooled, (iii) reduced by a scatter factor of 2.4, and (iv) compared with the original Safe Life from the full-scale test. This procedure indicated a life limit reduced to only 47% of the original Safe Life. This was adopted as an interim measure while the fleet recovery programme was carried out [89].

4.3.2 Fleet Recovery Programme

The fleet recovery programme was instigated because the interim reduced life limit meant that only 11 of the 69 aircraft in the remaining fleet would be airworthy by December 1991 [89]. The programme consisted of:

(1) Teardown, NDI and QF of detected fatigue cracks in another four wing spars. This meant that a total of ten spars, including those from A7-026, were investi-gated; a total of over 1000 bolt holes were examined and QF data were obtained for 103 cracks.

(2)   As in the case of the Safe Life reassessment:

- It was assumed that immediate in-service exponential FCG occurred from the measured initial discontinuities.
- FCG plots for the 'worst case' cracks were extrapolated to predicted failure points.
- The FCG life predictions were converted into Safe Life damage units, pooled and reduced by a scatter factor of 2.4, and then compared with the original Safe Life from the full-scale test. This resulted in a reassessed reduced life limit of 52% of the original Safe Life, close to the first estimate of 47%, and also unacceptable.

(3)   Considering several options: (i) safety-by-inspection; (ii) another full-scale fatigue test to obtain more confidence in the life limit; (iii) wing repairs; and (iv) wing replacements.

Only option (iv), wing replacements, proved to be feasible [89]. Thirty new wing sets were purchased to allow continued operation. In October 1994 a mid-air collision between two Macchi aircraft with replacement wings enabled assessing their build quality [89]. This was similar to that of the 'old' wings, and so the reduced life limit, rounded down to 50% of the original Safe Life was maintained. The fleet was withdrawn from service in 2001.

4.4   Lessons Learned

Broadly speaking, the MB-326H (S/N A7-076) accident changed the RAAF's structural integrity policy from the traditional Safe Life approach to Damage Tolerance. Specifically, this accident led to establishing structural integrity programmes, called ASIPs (Aircraft Structural Integrity Programmes), similar in scope to those of the USAF; and to the recognition that QF is essential to the structural integrity management of aircraft fleets [74,90] and the analysis of full-scale, component and specimen fatigue tests [65,66].

Two other important lessons are:

(1)   A follow-on full-scale fatigue test should have been done to validate the LOTEX replacement programme, including contemporary knowledge of the fleet usage (flight loads).
(2)   This accident demonstrated the need for a mid-life teardown of a high-time wing to validate the original Safe Life analysis.

Chapter 5
P&W 125B Engine Bearing (1994): 2-Stage Exponential FCG Analysis

5.1 Introduction

In February 1994 the right wing engine of a Fokker 50 aircraft failed during take-off, which was aborted. This incident was caused by fatigue failure of the No. 3 bearing, which allowed the low pressure impeller to move forward, resulting in secondary damage and subsequent engine failure [134]: the engine was internally destroyed. Figure 5.1 shows the aircraft and the type of engine, a Pratt & Whitney (P&W) 125B turboprop: the arrow points to the approximate internal location of the failed No. 3 bearing.

The failed bearing had experienced 5095 service hours (6425 engine starts) since new, and 530 service hours (585 engine starts) since overhaul by a P&W subsidiary in the Netherlands.

5.2 The Failed No. 3 Bearing: Macroscopic and Metallographic Examination

Disassembly of the failed No. 3 bearing revealed the following anomalies and information, see Fig. 5.2:

- Cage failure at the location of a considerably elongated ball pocket.
- Severe damage to 19 of the 20 balls, owing to overheating, deformation, cracking and wear. These balls had undamaged diameters ≈12 mm.
- One relatively undamaged ball with a diameter ≈11 mm. This ball had been in the elongated pocket where cage failure occurred.
- Cage and bearing overheating owing to contact between the cage and the outer race shoulders. The overheating most probably occurred after cage failure.

© The Author(s), under exclusive licence to Springer Nature B.V., part of Springer Nature 2019
R. Wanhill et al., *Fatigue Crack Growth Failure and Lifing Analyses*
for Metallic Aircraft Structures and Components, SpringerBriefs in Applied Sciences
and Technology, https://doi.org/10.1007/978-94-024-1675-6_5

Fig. 5.1 The incident aircraft, October 1993, and a view of the engine type (P&W 125B): the added arrow points to the approximate internal location of the failed No. 3 bearing. *Sources* Peter Gates Collection; P&W Canada Brochure 1988

- The cage microstructure was originally a fine-grained tempered martensite, typical of a bearing steel. Regions of untempered martensite had formed where the cage contacted the outer race, and also where a ball had rubbed against a spacer ligament between ball pockets.

5.3   Cage Failure and FCG Analysis

The failed cage fracture surfaces were typical for high-cycle—low stress fatigue, and were covered for more than 90% by FCG progression markings. Figure 5.3 gives a macroscopic view of one of the fracture surfaces, indicating the crack growth direction, which was from the cage outer surface. The inverse curvature of the progression markings near the inner surface is a consequence of their increasing proximity to this surface.

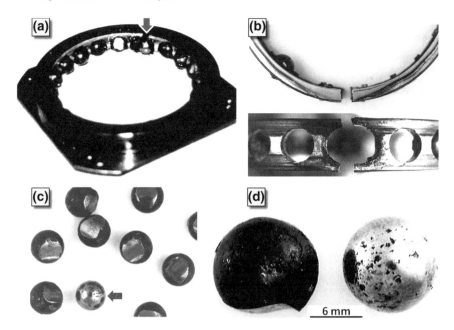

Fig. 5.2 Macroscopic examination of the No. 3 bearing: **a** the cage failure in situ (arrowed); **b** two views of the failed cage, note the elongated ball pocket; **c** some of the 20 balls, 19 of which showed severe damage and one (arrowed) with no damage; **d** details of a damaged ball and the undamaged ball, which was smaller in diameter ($\approx$11 mm) than all the other ($\approx$12 mm) balls. *Source* Barter SA, Defence Science and Technology Group DSTG, Melbourne, Australia

Many of the progression markings were similar and had a fairly uniform spacing. These most probably represented start–stop cycles of the engine. Although difficult to measure close to the cage outer surface, the early progression markings indicated that through-thickness cracking was already established at a crack depth of about 1.1 mm from the cage outer surface. This suggested fatigue nucleation from multiple origins, but severe edge rubbing prevented any observations of this. However, additional metallography on undamaged cage material showed that the cage had been etched before applying a silver coating, and that there were etch pits between 2 and 10 μm deep. These could probably have acted as fatigue-nucleating discontinuities.

5.3.1   Progression Marking QF Analysis

53 prominent progression markings were measured by optical fractography, covering crack depths of 1.9–5.3 mm, see Fig. 5.3. In the first instance, a plot of the QF measurements indicated approximately exponential FCG, which is shown compactly in Fig. 5.4. The complete diagram was derived with the following assumptions:

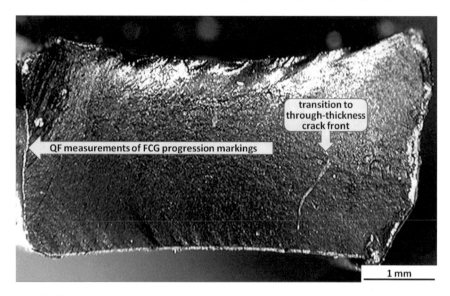

Fig. 5.3 Macroscopic view of one of the cage fracture surfaces: the inner and outer edges were very rubbed, but FCG progression markings were visible over much of the fracture surface

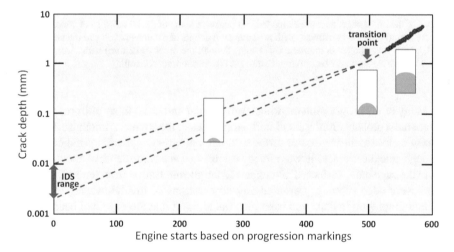

Fig. 5.4 Crack depths versus engine start–stop cycles for the fatigue crack in the steel bearing cage: IDS = initial defect size. The change from a semi-elliptical crack front to a through-thickness crack front is shown schematically in the diagram

(1) The prominent progression markings represented start–stop engine cycles.
(2) Multiple fatigue nucleation sites from initial defects (etch pits) in the size range 2–10 μm.
(3) 2-stage exponential FCG with a transition at a crack depth of 1.1 mm, owing to the change from a semi-elliptical crack front to a through-thickness crack front. This behaviour is consistent with more general knowledge about crack growth rate transitions from part-through-cracks to through-cracks.

5.4 Interpretation of the Results and Remedial Action

The bearing failure was due to insertion of an undersize ball during overhaul. This ball would have been unable to take loads, resulting in (i) abnormal out-of-radius loads on adjacent balls; (ii) abnormal tension loads on the cage; (iii) fatigue in the cage ligaments on either side of the undersize ball; (iv) cage failure and expansion to rub on the outer race; (v) rubbing-induced overheating, causing lubrication loss and complete failure of the bearing.

The remedial action was to install a new engine on the aircraft.

N.B: The OEM (P&W) originally wrote a letter to the Australian Transport Safety Bureau in which it was stated that '*The small ball is actually a correctly sized ball that had worn down to its present size through misuse during service.*' This statement was not supported by the failure analysis.

Chapter 6
EBA Example Lifing Assessment (2004): F/A-18 Horizontal Stabilator Spindles

6.1 Introduction

As discussed in Chap. 1, Sect. 1.2.4, the EBA is a framework for predicting the FCG lives of in-service aircraft structures subjected to relatively short and repeated blocks of VA loading sequences, i.e. load sequences representative for tactical aircraft. The EBA has been developed using the DST's extensive experience with VA full-scale fatigue testing in combination with specimen tests and QF of progression markings [67, 68].

A comprehensive flowchart of the EBA procedure is given in Fig. 6.1. Parts of the procedure may not be necessary, depending on (i) the availability of C_{va} and m_{va} values and (ii) whether it is decided to set $m_{va} = 2$, as recommended in Sect. 1.2.4. Also, either specimen or full-scale fatigue test QF data (or both) may be used, but only one source of data is actually required.

6.2 EBA Validation for the F/A-18 Horizontal Stabilator Spindles

The primary load-carrying attachments for the Boeing F/A-18 aircraft horizontal stabilators are AF1410 high strength steel spindles. Figure 6.2 shows the left-hand spindle location, between the two bulkheads Y645 and Y657.

In 1996 the DST started a unique full-scale fatigue test, designated FT46, of the aft fuselage and empennage. This test is discussed in more detail in [136], but here we note that the test combined buffet-induced dynamic loading and manoeuvre loading, with the above-mentioned characteristics of relatively short and repeated blocks of VA loading sequences. As a supplement to this test it was required to assess the FCG Damage Tolerance capability of the spindles. This was done via a notched coupon test programme, using the FT46 VA loading spectrum. The coupon tests were done

© The Author(s), under exclusive licence to Springer Nature B.V., part of Springer Nature 2019
R. Wanhill et al., *Fatigue Crack Growth Failure and Lifing Analyses
for Metallic Aircraft Structures and Components*, SpringerBriefs in Applied Sciences
and Technology, https://doi.org/10.1007/978-94-024-1675-6_6

at several reference stress levels to account for the stress distribution in the full-scale test articles (spindles). Deep focus optical QF was used to obtain the FCG data.

Before finalising the FCG Damage Tolerance assessment, several types of FCG models, including the EBA, were validated against the QF FCG data from the coupon

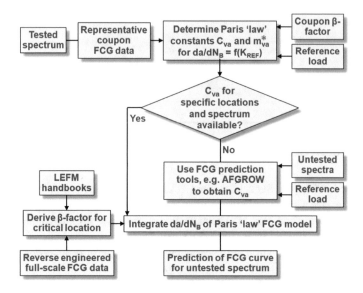

Fig. 6.1 Flowchart of the EBA procedure modified from [135]. N.B: ✳indicates that it is recommended to set $m_{va} = 2$, i.e. to assume exponential FCG, see Chap. 1, Sect. 1.2.4

Fig. 6.2 Location of the F/A-18 left-hand horizontal stabilator spindle. *Sources* RAAF and DST

tests [64]. The two best results are shown for comparison in Fig. 6.3, starting from a through-crack length of 2 mm. **N.B**: This crack size metric is important and equitable because the comparison avoids problems that CA-based cycle-by-cycle models can have in predicting short crack FCG. Nevertheless it is seen that the EBA provided better FCG predictions at the higher stress levels. These results, combined with the efficiency and relative steadiness of the predictions for longer cracks, led to selecting the EBA for the Damage Tolerance assessment [137]. This assessment is summarised in Sect. 6.3 as part of a total life assessment (safe fatigue life from the FT46 test + the EBA-predicted Damage Tolerance life).

6.3 EBA and Total Life Assessment for the F/A-18 Horizontal Stabilator Spindles

A full description of the EBA life assessment of the F/A-18 horizontal stabilator spindle is given in [137]. Here we present a summary of the procedure and results. As noted at the end of Sect. 6.2, the EBA life assessment was part of a total life assessment, consisting of the safe fatigue life determined from the FT46 full-scale test + the Damage Tolerance life predicted using the EBA.

The rationale for conducting this combined life assessment is as follows. Firstly, the FT46 full-scale fatigue test accumulated 23,090 simulated flight hours (SFH) by July 2002, at which time it was halted for economic reasons. The intention of this test was to provide the aft fuselage and empennage with a general Safe Life clearance of approximately 6000 flight hours. However, service loads on the spindles include highly variable dynamic buffet loading, leading to greater uncertainty in their fatigue lives. This meant that an unusually high scatter factor, up to 7, was required for Safe Life clearance of the spindles. On its own, the FT46 test duration (23,090 SFH) could not provide the necessary clearance for the spindles, even though no cracking was found. An additional Damage Tolerance life assessment using the EBA was done as an *effective* continuation of the FT46 test, whereby it was assumed that very small cracks in critical locations had not been detected. At the time, this was an innovation in the DST's testing and analysis evaluations.

The additional Damage Tolerance life assessment required the following steps:

(1) Determination of the critical locations on the spindles, using detailed 3D finite element method (FEM) modelling. Some results are depicted in Fig. 6.4, pointing out some of the critical locations: the upper and lower arm root fillets and two of the attachment areas.
(2) Validation of the FEM modelling by a peak strain survey on an actual spindle in an F/A-18 aft fuselage. The results for the upper and lower arm fillets were in good agreement, but some scatter for the attachment areas [137] resulted in the subsequent FCG analysis referring to a 'typical' attachment location.
(3) Determining β-factors for cracks growing into the upper and lower arm fillets and a typical attachment area. This was not a straightforward exercise, requiring

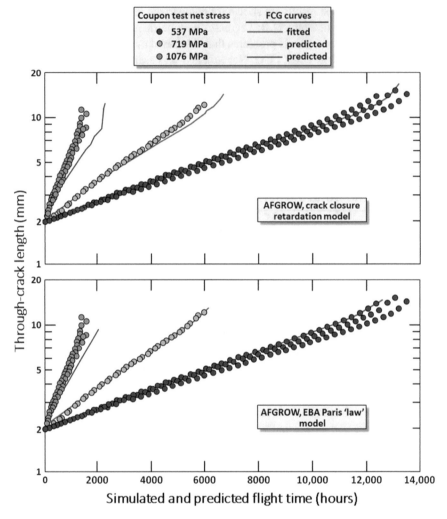

Fig. 6.3 Comparison of a CA-based cycle-by-cycle analytical crack closure model and the EBA Paris 'law' model against QF FCG data from AF1410 steel coupons. Each model was fitted to the low stress (537 MPa net) data and then used to predict the FCG at the higher stress levels: after [64]

(i) FEM and AFGROW modelling, and (ii) test validation (see point 4) of the FCG behaviour of an artificial defect (EDM notch) introduced into the upper aft attachment of the FT46 starboard spindle.

(4) Destructive inspection of the FT46 test starboard spindle at the upper and lower arm root fillets and attachment areas to measure any surface defects or discontinuities, and also conduct QF for the deliberately fatigue-cracked upper aft attachment of the FT46 starboard spindle. The maximum discontinuity depth

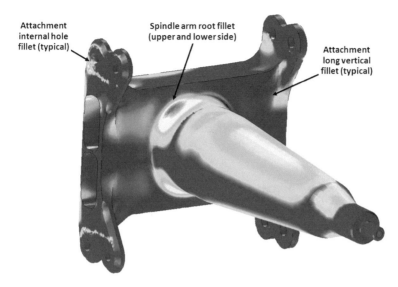

Fig. 6.4 Stress contour plots showing some of the most critical locations on the F/A-18 horizontal stabilator spindle [137]

found was 16 μm. This was rounded up to 20 μm to be used as the starting point for the FCG analyses.

(5) Conservatively neglecting any beneficial effects of residual stresses owing to standard production shot peening of the spindles.

(6) FCG analyses using AFGROW and the EBA Paris 'law' model, $da/dN_B = C_{va}(\sigma_{REF}\beta\sqrt{\pi a})^{m_{va}}$, calibrated by the QF results for the coupon tests, see Sect. 6.2, and the deliberately fatigue-cracked upper aft attachment of the FT46 starboard spindle.

The results of the FCG analyses, as part of a total life analysis, are shown in Fig. 6.5. The FCG curves extend to the maximum crack lengths that can sustain the required residual strength of the spindles. The estimated total lives of the lower arm root fillet and typical attachment locations were sufficiently long to meet the Safe Life clearance of 6000 flight hours. However, the estimated total life of the upper arm root fillet was not quite long enough: $38{,}790 \div 7 = 5541$. Therefore a separate EBA estimation was done with residual stress effects included. The results then showed that the upper arm root fillet reached an estimated total life of 49,790 SFH, easily meeting the Safe Life clearance requirement [137].

N.B: Following this demonstration of the EBA's practical utility, it was also used to validate the service life extension of the RAAF's General Dynamics F-111C fleet [88], which was finally retired in 2010.

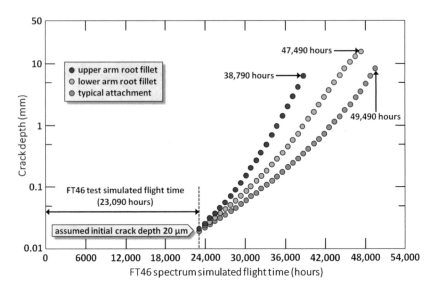

Fig. 6.5 Total life analysis results (FT46 test + EBA Damage Tolerance FCG curves) for the critical locations on the starboard F/A-18 horizontal stabilator spindle, excluding beneficial residual stress effects owing to shot peening: after [137]

Chapter 7
LCFLF Example Lifing Assessment (2004): F/A-18 Vertical Tail Attachment Stubs

7.1 Introduction

As mentioned in Chap. 6, in 1996 the DST began a unique full-scale fatigue test, FT46, of an F/A-18 aft fuselage and empennage assembly. This test article was especially acquired for testing. The test combined buffet-induced dynamic loading and manoeuvre loading at representative load frequencies, which was a formidable challenge for buffet loading [136]. Overall, the test load history had characteristics typical for tactical aircraft, i.e. relatively short and repeated blocks of VA loading sequences. Each block was equivalent to approximately 300 simulated flight hours (SFH).

As stated in Chap. 6, Sect. 6.3, the FT46 test article accumulated 23,090 SFH by July 2002, at which time the test was halted for economic reasons. The intention of this test was to provide the aft fuselage and empennage with a general Safe Life clearance of approximately 6000 flight hours after applying a scatter factor. However, inspections during testing revealed numerous *deficiencies*, mostly concerning cracking of the vertical tail attachment stubs. Figure 7.1 is a combined illustration of the attachment stub locations. The stubs were integral parts of AA7050-T7452 aluminium alloy forged formers that had been machined to final dimensions before chemical etching and applying an ion vapour deposition (IVD) aluminium coating for corrosion protection.

The stubs on the FT46 test article were 'second generation' OEM (Boeing) modifications to supposedly avoid nuisance (*non-critical*) cracking already observed on some service aircraft [136, 138]. The location of this type of cracking, which despite the modifications also occurred in the FT46 test, is illustrated in Fig. 7.2. More importantly, Fig. 7.2 shows the location of a stub flange crack that occurred early in the FT46 test. This is a *critical* type of cracking, and is the subject of the present Chapter.

Fig. 7.1 Locations of F/A-18 vertical tail attachment stubs. *Sources* RAAF and DST

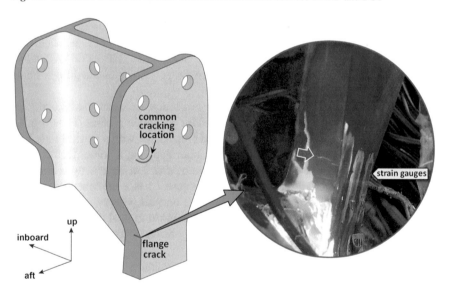

Fig. 7.2 The FT46 stub flange crack location and detection (red arrow) by dye penetrant under ultraviolet (UV) light. The common non-critical cracking location for some service aircraft (and also observed during the FT46 test) is also shown

7.2 The FT46 Stub Flange Crack

The FT46 stub flange crack was in the outboard stub flange of the starboard aft former. The crack was initially detected from strain gauge outputs after testing for 7846.3 SFH [136], and this was confirmed by dye penetrant inspection, see Fig. 7.2. The crack was excised, followed by:

(1) Examination of the affected flange area to ensure complete removal of the cracking.
(2) Contour blending and shot peening of the affected flange area before continuing the FT46 test.
(3) Fractographic examination of the removed crack, including QF.
(4) LCFLF assessment of the potential FCG life of the crack had it not been removed.

N.B: Detailed examination of FT46 cracks was part of a general RAAF/DST policy of removing them during testing (if necessary) and teardown, followed by QF to determine the fatigue-nucleating discontinuities and assist in analysing the FCG behaviour. In the following Sects. 7.3–7.5 the crack removed after 7846.3 SFH provides an example—one of many [65,66]—whereby the LCFLF approach to analysing the FCG behaviour and enabling lifing assessments has been found suitable for RAAF aircraft [85].

7.3 Fractography of the FT46 Stub Flange Crack

Figure 7.3 is an overview of one of the fracture surfaces from the excised and broken-open FT46 stub flange fatigue crack. This was wider and deeper than expected, so that part of it had been inadvertently cut away: the curved red line is an estimate of the extent of the crack before removing it. Fatigue nucleated from many etch pits that were due to chemical etching before IVD coating. The main fatigue origin was located at the position shown in Fig. 7.4a, which also shows the IVD coating layer. This fatigue origin was used as the starting point for QF, which followed the trajectory indicated by the blue dashed line in Fig. 7.3.

Progression markings identifying complete blocks of simulated flights, as in Fig. 7.4b, enabled deep focus optical QF over the entire trajectory of cracking. The data are shown in Fig. 7.5, starting from an EPS of 5 μm. This is smaller than the DST-determined average EPS (9 μm) for etch pits, see Table 3.1 in Chap. 3. A possible reason is that many adjacent etch pits acting as fatigue-nucleating discontinuities would avoid the need for a larger etch pit as the primary fatigue origin.

Fig. 7.3 Optical fractograph of the excised FT46 stub flange fatigue crack, showing the estimated extent of the crack (curved red line) and the QF trajectory (blue dashed line)

(a)   **(b)**

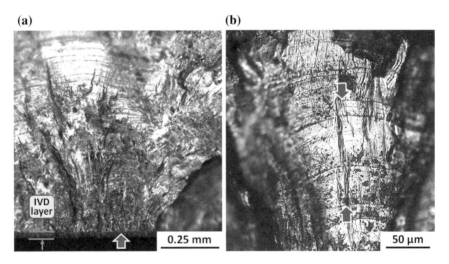

Fig. 7.4 Optical fractograph details of the excised FT46 stub flange fatigue crack: **a** main fatigue origin, also showing the IVD coating layer; **b** an example of a complete block of simulated flights, equivalent to about 300 simulated flight hours (SFH)

7.4   LCFLF Life Assessment for the FT46 Stub Flange Crack

Figure 7.5 shows that the QF-obtained FCG data for the FT46 stub flange crack follow an exponential trend. This enabled using the LCFLF methodology, see Table 1.1 in Chap 1, for life assessment. A straight line drawn from the EPS to the critical crack length, a_{crit}, represents the estimated FCG life, about 10,600 SFH, if the crack had not been removed. This estimate represents the total life to failure of this type of crack in a typical fleet aircraft, since cracking may be assumed to begin almost immediately when an aircraft enters service [65, 66].

N.B: It was already mentioned in Chap. 6, Sect. 6.3, that the highly variable dynamic buffet loading in the FT46 test required an unusually high scatter factor for Safe Life clearance. This scatter factor was between 5 and 7, which when applied to the estimated FCG life of the stub flange crack gives only 1514–2120 SFH, i.e. much less than the target 6000 flight hours. This potential problem was resolved by introducing another stub modification for any fleet aircraft exceeding approximately 1500 flight hours.

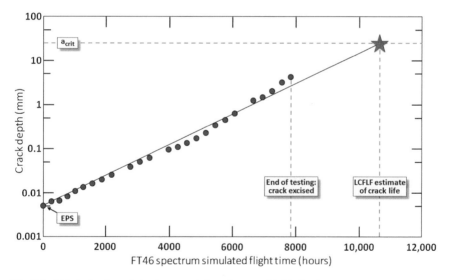

Fig. 7.5 QF-obtained crack depth versus full-scale test FCG life for the excised FT46 stub flange fatigue crack, and the LCFLF estimate of the total FCG life if the crack had not been removed

Chapter 8
Cubic Rule Life Prediction Examples

8.1 Introduction

As discussed in Chap. 1, Sect. 1.2.5.2, the cubic rule is a special category of the LCFLF. The cubic rule's premise is that the FCGR ratio for two tests using the same load spectrum, but at two different reference stress levels, is given by

$$(da/dN)_1/(da/dN)_2 = (\sigma_{REF.1}/\sigma_{REF.2})^{\alpha} \tag{8.1}$$

where $\alpha = 3$, hence the rule's designation. This rule implies that the lead crack growth rates obtained at one reference stress level may be used to predict the lead crack growth rates for a different reference stress level.

In practical terms, the cubic rule can increase the efficient use of QF-derived FCG data for life predictions. This is because the rule enables conservative (i.e. lead crack) predictions of the FCG life at different reference stress levels. **N.B**: For FCG life predictions it is always necessary to estimate EPS values for use in specific equations of the type $a = a_0 exp(\lambda N)$: see Eq. (1.14) in Chap. 1, Sect. 1.2.5.

Important applications include: (i) fatigue lifing of structural repairs [92], which may involve (limited) changes in local stress levels; and (ii) efficiently using QF-derived lead crack FCGR data from one location to predict the lead crack behaviour at other locations that experience the same load spectrum but at different stress levels.

An extensive evaluation of the cubic rule is given in [91], from which two examples are presented in this Chapter. These examples have been chosen to illustrate two methods (Methods 1 and 2) of analysis and the cubic rule's versatility. One example concerns a tactical aircraft, the McDonnell Douglas F4E, and the other is for a maritime patrol aircraft, the Lockheed Martin P3C.

8.2   McDonnell Douglas F4E Spectrum Coupon Test Data

Figure 8.1 presents $40\times$ optical microscopy FCG data obtained from AA7075-T6511 aluminium alloy coupons tested under flight simulation loading that was derived from a McDonnell Douglas F4E wing bending moment spectrum [139]. The coupons were 12.7 mm thick and contained central holes notched on one side to start corner cracks. There were two reference stress levels: 207 and 248 MPa, representing a baseline (0%) and 20% stress increase. In both cases the test data show a bilinear exponential FCG trend, with initial crack growth having a steeper slope.

The cubic rule was applied to the 207 MPa baseline data, using the steeper initial slope to predict FCG at the 248 MPa stress level [78]. This is cubic rule Method 1 [91]. Figure 8.1 shows that the prediction agrees conservatively with the 248 MPa test data. More details are given in [78,91], but here we note that the FCG data were extrapolated back to zero simulated flight hours (SFH = 0) to estimate the EPS *for both sets of data*. In other words, the prediction involved a separate EPS estimation for the 248 MPa test data. This is not the normal situation in practice, where predictions are intended to be made from a single set of baseline lead crack data.

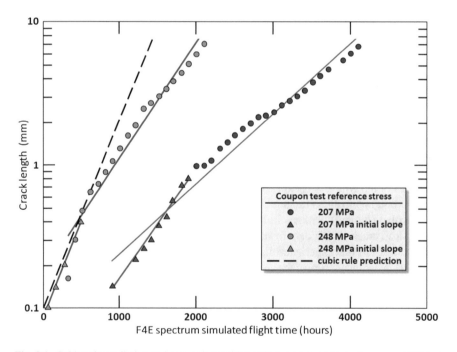

Fig. 8.1 Cubic rule prediction and comparison with notched coupon optical microscopy FCG test data obtained under F4E flight simulation loading [139]. The cubic rule used the 207 MPa initial slope data (Method 1) to predict FCG at the 20% higher stress level, 248 MPa. The prediction is suitably conservative. After [91]

8.3 Lockheed Martin P3C Spectrum Coupon Test Data

Figure 8.2 presents data envelopes for automated potential drop (PD) measurements of FCG obtained from AA7075-T6 aluminium alloy coupons tested under Lockheed Martin P3C flight simulation loading at the DST. The coupons were approximately 2 mm thick and contained central holes with starter notch pre-cracks on either side [91].

There were three reference stress levels representing a baseline (0%) and 10 and 20% stress increases: the baseline reference stress was a peak stress of 124.8 MPa [91]. An exponential regression fit was made to the baseline data, extrapolating back to SFH = 0 to estimate the fit's EPS value. Then this EPS and the cubic rule were applied to this average fit to predict FCG at the higher stress levels. This is cubic rule Method 2 [91]. Figure 8.2 shows that the predictions agree conservatively with the test data.

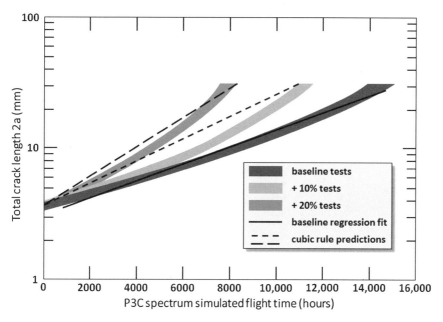

Fig. 8.2 Cubic rule predictions and comparisons with pre-cracked coupon FCG test data obtained under P3C flight simulation loading. The cubic rule used the baseline data regression line (Method 2) to predict FCG at the 10 and 20% higher stress levels. The predictions are suitably conservative

8.4 Summary

Since the cubic rule is derived from the LCFLF, the rule is intended to be used for lead cracks. This is important, since it maximises the potential for obtaining good but conservative predictions. The cubic rule's versatility with respect to different types of load spectra and aerospace structural alloys has been amply demonstrated [91], and two examples have been included in the present Chapter.

The RAAF has approved the use of the cubic rule by certified contractors for maintenance and structural repairs of several types of aircraft, including the F/A-18 and P3C fleets.

Chapter 9
Fokker 100 Fuselage Test: Lap Joints Exponential FCG Analysis

9.1 Introduction

In the mid-1980s Fokker Aircraft began a full-scale fatigue test of the fuselage and wings of the F100 aircraft [140]. During the test, which ran to 126,250 simulated flights, fatigue cracking occurred along one of the pressure cabin longitudinal lap splices, see Fig. 9.1. The lap slice had been assembled by adhesive bonding and riveting, and cracking extended over several frame bays that had poor adhesive bonds. Figure 9.2 show details of the lap splice configuration and the type of fatigue cracking. The cracking occurred along the rows of fastener holes, and at each hole, thereby constituting what is called Multiple Site fatigue Damage (MSD).

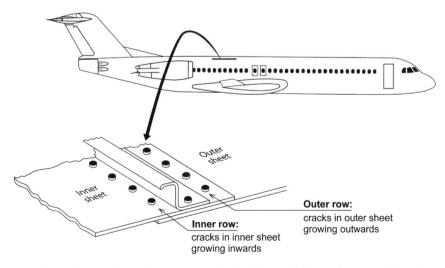

Fig. 9.1 Location of full-scale fatigue test cracking in Fokker 100 fuselage longitudinal lap splices

R. Wanhill et al., *Fatigue Crack Growth Failure and Lifing Analyses for Metallic Aircraft Structures and Components*, SpringerBriefs in Applied Sciences and Technology, https://doi.org/10.1007/978-94-024-1675-6_9

(a)

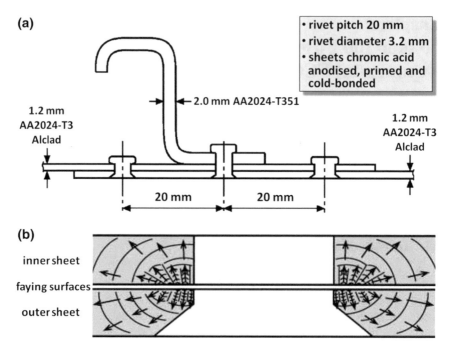

- rivet pitch 20 mm
- rivet diameter 3.2 mm
- sheets chromic acid
  anodised, primed and
  cold-bonded

2.0 mm AA2024-T351

1.2 mm
AA2024-T3
Alclad

1.2 mm
AA2024-T3
Alclad

20 mm   20 mm

(b)

inner sheet

faying surfaces

outer sheet

Fig. 9.2 Schematics of **a** the lap splice configuration and **b** example of fatigue crack nucleation and growth at one of the fastener holes in the cracked lap splice [72, 95]

After completion of the full-scale test the lap splice cracks were fractographically examined by SEM to determine the FCG behaviour at a number of holes. In particular, the intention was to estimate when the cracks became through-thickness and long enough (about 2 mm beyond the fastener hole bores) to be detectable on either the outer sheet external surface or the inner sheet internal surface.

9.2 FCG Analysis of the Cracked Lap Splice [71, 72, 95]

QF showed ductile fatigue striations representing essentially CA loading corresponding to once-per-flight cabin pressurization cycles, and without progression markings. Thus it was necessary to measure the striation spacings to determine the FCG behaviour. The measurements were done in the sheet thickness (transverse) direction and longitudinally, see Fig. 9.3a. Note that there were generally multiple fatigue nucleation sites, covering a crack length defined as a_0.

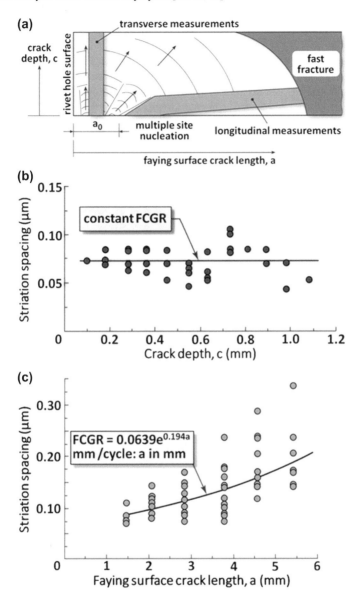

Fig. 9.3 **a** Schematic of the QF measurements; **b** and **c** representative examples of the QF measurements [72, 95]

Figures 9.3b and c show representative examples of the QF measurements. Despite the scatter, the data suggest on average a reasonably constant FCGR in the transverse direction and approximately exponential FCG in the longitudinal direction. Based on these results and the crack nucleation and growth characteristics illustrated in Fig. 9.3a, the original investigator [71] proposed (but did not describe) a model for the lap splice MSD, see Fig. 9.4. This model accounts for the trends in Fig. 9.3, but with an additional observation and three main assumptions:

(1) Observation: Figs. 9.3b and c show that the transverse and longitudinal striation spacings, and hence the crack growth rates, were similar for crack dimensions less than twice the sheet thickness (1.2 mm, see Fig. 9.2a).

(2) Assumptions: (i) constant FCGR in the transverse direction, equal to the initial FCGR in the longitudinal direction, i.e. $dc/dN = Ae^{Ba_0}$; (ii) crack depth $c = 0$ at a_0; (iii) quarter-circular crack fronts in the transition from transverse to longitudinal crack growth.

These assumptions are convenient but not essential. The model *could* be used for non-constant dc/dN. More directly, Fig. 9.5 gives examples of the model's actual use for non-countersunk and countersunk sheets: in the latter case for crack nucleation at a rivet hole corner, i.e. $a_0 = 0$ [71]. In both cases the model predicted that the cracks would have become through-thickness at about 95,000 simulated flights, and at maximum lengths of 2–3 mm. Whether this type of information was used in certification of the aircraft, including inspection schedules, has not been reported in the open literature.

● **From QF of striation spacings**

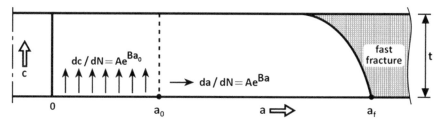

● a_0, a_f and N_f are known. Calculate N_0 from:

$$N_f - N_0 = \frac{1}{AB}\left(e^{-Ba_0} - e^{-Ba_f}\right)$$

● Calculate intermediate values of a for given values of N:

$$a_{int} = -\frac{1}{B}\ln\left[e^{-Ba_0} - AB\left(N_{int} - N_0\right)\right]$$

● For each a_{int} calculate c_{int} from:

$$c_{int} = \left(N_{int} - N_0\right)Ae^{Ba_0}$$

● Construct crack fronts as follows:

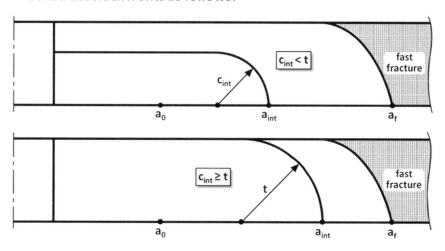

Fig. 9.4 Eijkhout's FCG model [71, 72, 95] for lap splice MSD, shown here for a non-countersunk sheet

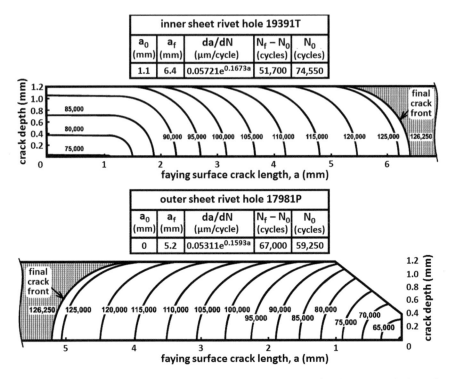

Fig. 9.5 Examples of the use of Eijkhout's model for non-countersunk and countersunk sheets: in the latter case for crack nucleation at a rivet hole corner, i.e. $a_0 = 0$

Chapter 10
Sikorsky S-61N Rotor Blade (1974): Exponential FCGR Analysis

10.1 Introduction

In May 1974 a Sikorsky S-61N helicopter, operated by KLM Noordzee Helikopters, crashed into the North Sea owing to failure of a main rotor blade [73, 93]. Figure 10.1 shows the wreckage during recovery. The indicated blade 3 failed by fatigue and fast fracture just before the crash, whereas the other blades broke during the crash. Figure 10.2 shows (i) the recovered fracture surface of blade 3, which consisted of a hollow AA6061-T6 aluminium alloy spar bonded to an aerodynamically shaped aluminium-skinned trailing edge pocket; and (ii) a summary of the optical and SEM fractographically determined failure sequence [73]. The first phase in this sequence was high-cycle fatigue nucleating from corrosion pits: corrosion had probably occurred because of *local* disbonding of the spar and pocket.

During service the blade was pressurized with nitrogen gas as part of an inspection system (pressure loss) intended to detect early through-thickness fatigue cracking of the spar. The inspection system consisted of pressure gauges installed on each blade near the rotor hub. Simulated spar/pocket bonds subjected to FCG (spar) and leak tests showed that the inspection system might not detect a spar crack until it had grown beyond the closest pocket edge, since the pocket skin need not have cracked nor fully disbonded from the spar [141]. This meant that the maximum amount of FCG life available for crack detection could have corresponded only to the phases 4 and 5 shown in Fig. 10.2. This conservative view is also justified because the pressure gauges could be checked only when the rotor hub was stationary, i.e. when the helicopter was on the ground or a platform; and then a crack that had not resulted in a detectable pressure loss would be closed by compression (self-weight) loading of the underside of the spar.

R. Wanhill et al., *Fatigue Crack Growth Failure and Lifing Analyses for Metallic Aircraft Structures and Components*, SpringerBriefs in Applied Sciences and Technology, https://doi.org/10.1007/978-94-024-1675-6_10

Fig. 10.1 S-61N helicopter PH-NZC during recovery: blade 3 failed by fatigue and fast fracture. *Source* RLD (Netherlands National Aviation Service)

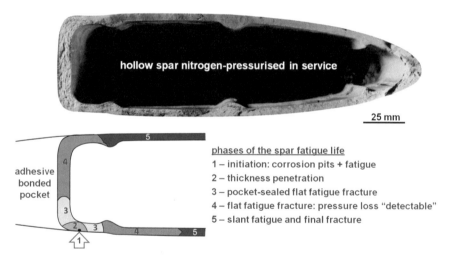

Fig. 10.2 Blade 3 fracture surface: AA6061-T6 aluminium alloy extruded hollow spar

10.2   FCG Analysis for the Crashed Helicopter

Besides the failure sequence indicated in Fig. 10.2, TEM and SEM fractography of phases 4 and 5 along the fracture surface at the lower side of the spar showed only ductile fatigue striations, mostly larger than 1 μm, until slant fracture, i.e. there was no evidence of corrosion-enhanced FCG. The striation characteristics suggested that the fatigue loading was essentially CA and the FCG rates were high.

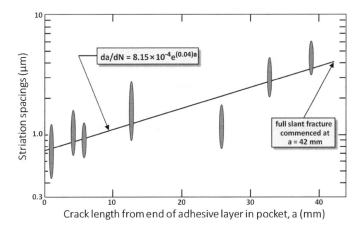

Fig. 10.3 Data envelopes for fatigue striation spacings, five measurements per position, for phase 4 FCG along the lower side of the blade 3 spar. A best fit exponential FCGR equation is also shown. After [73, 93]: **N.B**: This diagram is also shown as Fig. 1.9 in Chap. 1

Figure 10.3 presents the fatigue striation data envelopes for phase 4, with a best fit exponential FCGR equation. Integration of this equation gave an FCG life of 24,957 cycles for phase 4. Assuming a constant rotor speed of 203 rpm (Sikorsky) and that FCG occurred with a frequency of one striation per rotation, this FCG life corresponds to an estimated flight time of 2 h. Lower and upper bound flight times of 1.5 and 3.75 h were also estimated [93].

The estimates of phase 4 FCG lifetimes, corresponding to almost the total flight time for crack detection, are short. In fact, any estimate less than 3 h, representing a relatively long flight, is dangerously short, implying that blade failure could occur after a satisfactory pre-flight pressure gauge inspection. Although not proven, this seems the most plausible final contributing cause for the accident besides spar/pocket local disbonding and subsequent corrosion-induced fatigue nucleation.

10.2.1 Prediction of Detectable FCG Lives

The short FCG life estimates discussed in Sect. 10.2 prompted an investigation to determine how the detectable FCG life varied with indicated air speed (IAS). The IAS on the last flight of the crashed helicopter was assumed to be either 120 or 130 knots, which were representative cruise speeds. Predictions of detectable FCG lives were obtained from the best fit exponential FCGR equation in Fig. 10.3, an FCG design plot [142], and rotor blade stress measurements [143].

An analytical procedure for making these predictions is given in Table 10.1 and shown as a flowchart in Fig. 10.4. Note that as mentioned in Chap. 1, Sect. 1.2.5.3, the procedure used an exponential FCGR equation and a 'Paris law' equation, underpinning the pragmatic approach of 'whatever works best'.

Table 10.1 Analytical procedure for predicting detectable FCG life dependence on IAS

- Exponential best fit to QF plot, Fig. 10.3:
  - $da/dN = 8.15 \times 10^{-4} e^{0.04a}$, with a in mm (10.1)
  - integrate Eq. (10.1) from pocket edge (a_p) to full slant ($a_{fs} = a_p + 42$ mm) to obtain FCG life: $N = 24{,}957$ cycles

- LEFM 'Paris Law' derivation from Sikorsky design plot, see Fig. 10.4:
  - best fit striation spacings give $\Delta K_{ap} = 24.3$ MPa$\sqrt{m}$ and $\Delta K_{fs} = 33.2$ MPa$\sqrt{m}$
  - use these striation spacings (as da/dN values) and ΔK values to obtain a reference 'Paris law':
  $$da/dN = C(\Delta K)^m = 2.365 \times 10^{-15}(\Delta S\sqrt{\pi a})^6 \ (10.2)$$

- Integration of Eq. (10.2) from a_p to a_{fs}:
  $$N = \frac{1}{C(\Delta S\sqrt{\pi})^m} \cdot \frac{1}{(m/2)-1}\left[1/a_p^{(m/2)-1} - 1/a_{fs}^{(m/2)-1}\right] \ (10.3a)$$
  $$= \frac{6.818 \times 10^{12}}{(\Delta S)^6}\left[1/a_p^2 - 1/a_{fs}^2\right] \ (10.3b)$$
  where a_p and a_{fs} are *effective* crack lengths in metres

- Derivation of FCG lives for different indicated air speeds (IAS):
  - main rotor blade stress ranges as f(IAS) obtained from 100% design speed, see Fig. 10.4
  - assume PH-NZC final flight IAS was either 130 knots or 120 knots
  - use Eq. (10.3b) iteratively to obtain a_p and a_{fs} for $N = 24{,}957$ cycles and $\Delta S_{130 \text{ knots}} = 77.2$ MPa and $\Delta S_{120 \text{ knots}} = 70.3$ MPa
  130 knots: $a_p = 0.032$ m, $a_{fs} = 0.074$ m; 120 knots: $a_p = 0.041$ m, $a_{fs} = 0.083$ m
  - calculate N for lower IAS using these two sets of a_p and a_{fs} values
  - convert N into flight hours assuming a constant rotor speed of 203 rpm and that FCG occurred with a frequency of one cycle per revolution

10.3 Results and Remedial Actions (Lessons Learned)

Figure 10.5 shows firstly the analysis results, and secondly a restriction on the maximum speed after an in-flight pressure loss indication. It is seen that restricting the IAS after a pressure loss indication increases the detectable FCG life. This is due to a reduction in the blade vibratory stresses, as indicated by the design speed diagram in Fig. 10.4.

More specifically, a reduction in IAS to 90 knots increases the detectable FCG life to beyond 12 h, i.e. more than four times the maximum flight time of about 3 h. This result was coupled to Sikorsky's development of cockpit *in-flight* pressure indicators, such that any indication of blade pressure loss during flight would be immediately followed by a reduction in IAS to 90 knots. This measure was implemented for the S-61N helicopter fleet.

Furthermore, previous maintenance practice on the main rotor blades was upgraded. Until the time of the S-61N accident a certain amount of disbonding between the spars and pockets was allowed. Afterwards, owing to the primary cause of the accident, namely corrosion-induced fatigue nucleation, a directive was issued stating that any detected disbonds were to be sealed immediately and repaired within 2 months.

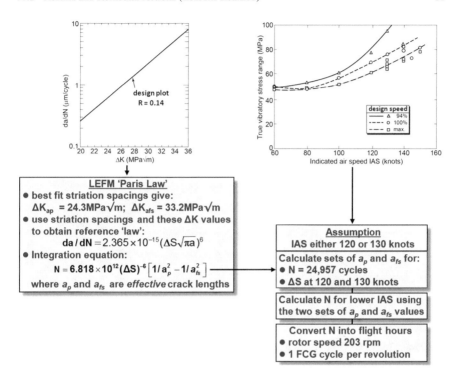

Fig. 10.4 Flowchart for predicting detectable FCG life dependence on IAS

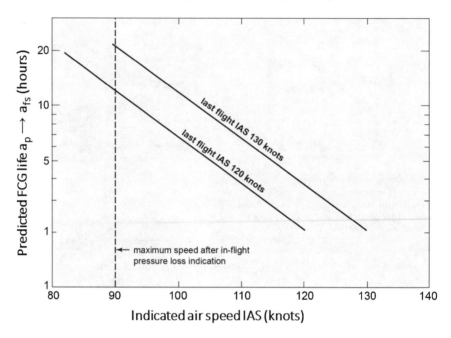

Fig. 10.5 Predicted FCG lives from pocket edge, a_p, to full slant fracture, a_{fs}, as functions of IAS

Chapter 11
Westland Lynx Rotor Hub (1998): Progression Marking LEFM Analysis

11.1 Introduction

In November 1998 a Westland Lynx helicopter, operated by the Royal Netherlands Navy (RNLN), lost a rotor blade and then the rotor head just before take-off from the RNLN base at Den Helder, the Netherlands. Figure 11.1a shows the helicopter at an earlier date with an indication of the failure location. The failure was caused by fatigue and fracture of the so-called yellow arm of the rotor hub. Figure 11.1b shows the outboard part of the fracture, which was across an elliptical plane section. The rotor hub, with GKN Westland (GKNW) designation M323, failed after 3591.9 service hours, well below the overall design safe retirement life of 5000 h minimum and 8600 h at the failure location [144].

The yellow arm was one of four arms of the rotor hub, which was a monolithic Ti-6Al-4V forging ($\alpha+\beta$) processed and heat-treated to obtain a microstructure consisting of primary α and transformed β [119, 145]. An extensive investigation found no evidence for failure due to poor material processing (microstructural anomalies) and properties. Also, the hub final machining, surface finishing (shot peening) and dimensions were within the specifications [119, 145], and no service damage was found.

Attention then concentrated on fractography of the failure location in the yellow arm, followed by similar fractography of fatigue fracture surfaces from (i) a full-scale, 4-stress-level variable amplitude (VA) test on another arm, with hub designation M6; and (ii) representatively shot-peened specimens made from the M326 yellow arm and tested under fully reversed stressing ($R = -1$) by GKNW. Finally, the fractographic and microstructural results and test stress levels were used to estimate the fatigue stress level(s) responsible for the service failure.

R. Wanhill et al., *Fatigue Crack Growth Failure and Lifing Analyses for Metallic Aircraft Structures and Components*, SpringerBriefs in Applied Sciences and Technology, https://doi.org/10.1007/978-94-024-1675-6_11

Fig. 11.1 a Lynx helicopter showing location of later rotor hub failure; **b** outboard fracture surface of the hub 'yellow' arm before detailed examination. *Source* RNLN (Royal Netherlands Navy)

11.2   Fractography of Early Fatigue Cracking in the M326 'Yellow' Arm

11.2.1   Progression Markings

Optical and low-magnification SEM fractography of the M326 yellow arm fracture surfaces revealed several progression markings close to the fatigue origin, see the schematic in Fig. 11.2. This schematic is correctly proportioned with respect to the phases of fatigue nucleation and early crack growth. The nucleation (phase 1) and phase 2 crack growth occurred subsurface, effectively *in vacuo*, owing at least in part (if not entirely) to shot-peening the surface of the arm. When the crack broke through to the surface (phase 3) there was a subtle change in fracture topography [119,145].

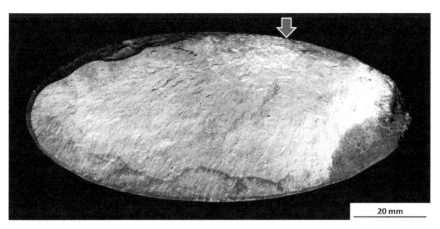

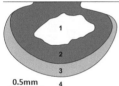

phases of the 'yellow' arm fatigue life
1 – highly faceted subsurface fatigue nucleation *in vacuo*
2 – rough early crack growth *in vacuo*
3 – rough early crack growth in air
4 – less rough crack growth in air

Fig. 11.2 Outboard fracture surface of the M326 'yellow' arm with a schematic of early FCG from the fatigue nucleation area (arrowed)

More importantly, see Sect. 11.4, the transition from phase 3 to phase 4 was marked by a *decrease* in fracture surface roughness. This transition was designated the fracture roughness transition boundary (FRTB).

11.2.2   Detailed Fractography

SEM fractography of the areas covered by phases 1–4 showed that the fracture surfaces were a combination of cleavage-like fracture through α grains and complex microtopographies associated with the transformed β grains [119, 145]. In fact, the detailed fracture surface appearances broadly resembled those shown in Figs. 3.1f and g in Chap. 3. Easily recognisable fatigue striations suitable for QF were not observed.

11.3   Fractography of Early Fatigue Cracking in the M6 Arm and M326 Specimens

Optical and low-magnification SEM fractography of the M6 arm and M326 specimen fracture surfaces showed progression markings analogous to those on the fracture surfaces of the failed M326 yellow arm. The M6 arm FRTB showed the same effect as the M326 yellow arm FRTB, i.e. a *decrease* in fracture surface roughness when crossing from phase 3 to phase 4. However, the M326 specimen FRTBs showed a relative *increase* in roughness owing to the phase 3 fracture surfaces being flattened by severe contact between them. This reverse trend was attributed to the fatigue test condition of fully reversed cyclic stresses, whereby originally rougher fracture surfaces would be liable to more severe contact between them [119, 145].

11.4   FRTB + LEFM Estimations of Damaging Service Fatigue Stress Levels

The key to this part of the service failure investigation was the FRTB, which was known to occur at reasonably well-defined ΔK values under both CA and VA loading [145]. In detail, the FRTB ΔK values depend mainly on the titanium alloy primary α grain size, cyclic yield stress, and fatigue stress ratio, R [146]. A complete description of the methods for estimating the damaging fatigue stress level(s) that resulted in failure of the yellow arm is given in [119]. These methods are outlined in Table 11.1 and shown as a flowchart in Fig. 11.3. Estimates of the damaging alternating stress levels (maximum frequently occurring alternating stresses) ranged from 332–484 MPa. The most likely value was about 370 MPa, which is higher than the alternating stress endurance limit for the yellow arm failure location: 299–342 MPa [149].

11.5   Measurements of Service Loads and Fatigue Analysis by GKNW

The service load conditions that could be relevant to the Lynx accident were investigated and measured by GKNW. A condition designated *minimum pitch on ground* (MPOG) was found capable of including fatigue stresses above the rotor hub endurance limit. This load condition had not been included in the original fatigue safe life (retirement) analysis of the rotor hub. GKNW redid the life analysis using the well-known Pålmgren–Miner rule, and concluded that *up to 70% of the fatigue damage could have been contributed by the MPOG loads* [150]. Although the Pålmgren–Miner rule's reliability is questionable, this emphatic result suggests that MPOG loads were primarily responsible for the yellow arm failure.

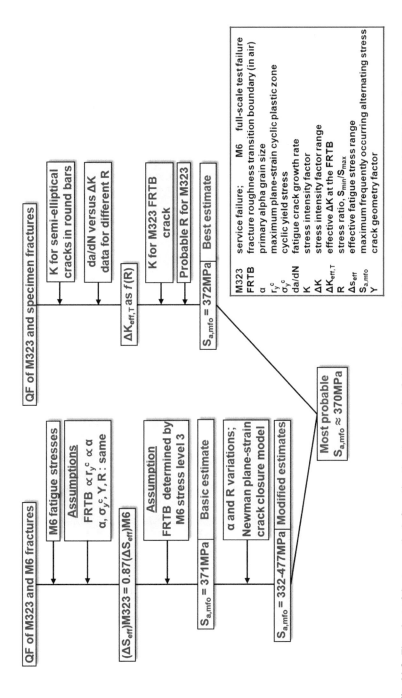

Fig. 11.3 Flowchart of the two methods for estimating the damaging fatigue stress level(s) that resulted in failure of the M326 'yellow' arm

Table 11.1 Outline of the two methods for estimating the damaging fatigue stress level(s) that resulted in failure of the M326 'yellow' arm

• QF mapping the FRTBs for the service, full-scale and specimen test failures

• **Method 1**: comparison of service and full-scale test failures using 6 assumptions:
(1) FRTBs correspond to the maximum plane-strain cyclic plastic zone becoming equal to the primary α grain size [146]
(2) Full-scale test FRTB determined by level 3 (of 4 levels) fatigue stresses [119]
(3+4) Same cyclic yield stress and crack geometry factor
(5+6) Same or different primary α grain size and fatigue stress ratio, R

• **Method 2**: comparison of service and specimen test failures using 4 assumptions:
(1) FRTBs due to the same mechanism
(2) FRTB ΔK_{eff} as $f(R)$ valid for the service and specimen failures [119]
(3) Specimen FRTB stress intensity factors derivable from Murakami's Handbook [147]
(4) Service failure FRTB stress intensity factor derivable from the expression for a semi-circular surface crack in a thick plate [148]

• Estimates of the damaging alternating stresses in the yellow arm at the failure location ranged from 332–477 MPa (Method 1) and 372–484 MPa (Method 2)

Support for this suggestion is that the MPOG loads contributed about 11% of the total number of cycles. This percentage is similar to that of the second-highest (level 3) stress level occurrences (12.1%) in the full-scale fatigue test [119]; and Table 11.1 and Fig. 11.3 show that choosing stress level 3 is a key assumption for Method 1. **N.B**: this is not a circular argument, since the Method 1 estimate and the GKNW calculations were independent and completely different.

11.6 Causes of Failure and Remedial Action

The yellow arm failure was due to too-high dynamic stresses on the rotor hub, in turn the result of inadequate fatigue design, whereby the MPOG loads were not included in the analysis. The remedial action for other Lynx helicopters was a new design multi-element damage tolerant rotor hub made from the β titanium alloy Ti-10V-2Fe-3Al, which has better fatigue properties than Ti-6Al-4V [151]. The new design rotor hubs were gradually introduced into the RNLN fleet.

Chapter 12
Closing Remarks

In keeping with the aims of SpringerBriefs, this book is a summary of some currently available methods for fatigue crack growth (FCG) analysis and lifing of metallic aircraft structures and components. These methods are drawn from the authors' experiences going back, in one case, to the 1970s.

A historical review of FCG parameters, models, and analysis methods provides the background for briefly discussing the importance of fatigue for aircraft structures and FCG analysis techniques. In the authors' opinion it is essential that fractography, in particular quantitative fractography (QF), be used as far as possible in service failure FCG analyses *if they are to be reliable*. QF is also an important, and often essential, addition to FCG analyses of full-scale, component and specimen test results.

Several case histories from the authors' experience have been selected to illustrate most of the FCG analysis methods discussed in the historical review, with an additional special case in Chap. 11. The case histories are very different, but all of them include QF as an essential contribution. Notwithstanding this requirement, the selection of FCG models is characterized by variety and pragmatism in approximating reality, based on 'whatever works best' and was available at the time.

Neither the selection and use of FCG analysis methods nor the use of QF is an easy task. We hope that this book, and the referenced literature, will be helpful in accomplishing these tasks during other aircraft service failure and lifing analyses.

R. Wanhill et al., *Fatigue Crack Growth Failure and Lifing Analyses for Metallic Aircraft Structures and Components*, SpringerBriefs in Applied Sciences and Technology, https://doi.org/10.1007/978-94-024-1675-6_12

References

1. Shanley FR (1952) A theory of fatigue based on unbonding during reversed slip, Report P-350, November 1952. The Rand Corporation, Santa Monica, CA, USA
2. Head AK (1953) The growth of fatigue cracks. Philos Mag 44:925–938
3. Frost NE, Dugdale DS (1958) The propagation of fatigue cracks in sheet specimens. J Mech Phys Solids 6(2):92–110
4. McEvily AJ Jr, Illg W (1958) The rate of fatigue-crack propagation in two aluminum alloys, NACA Technical Note 4394, September 1958. National Advisory Committee for Aeronautics, Washington, DC, USA
5. Illg W, McEvily AJ Jr (1959) The rate of fatigue-crack propagation for two aluminum alloys under completely reversed loading, October 1959, NASA Technical Note D-52. National Aeronautics and Space Administration, Washington, DC, USA
6. Liu HW (1959) Crack propagation in thin metal sheet under repeated loading, PhD Thesis, University of Illinois, Urbana, IL 61801, USA
7. Forsyth PJE (1957) The application of 'fractography' to fatigue failure investigations, Technical Note No: MET 257, March 1957. Royal Aircraft Establishment, Farnborough, UK
8. Ryder DA (1958) Some quantitative information obtained from the examination of fatigue fracture surfaces, Technical Note No: MET 288, September 1958. Royal Aircraft Establishment, Farnborough, UK
9. Paris PC, Gomez MP, Anderson WE (1962) A rational analytic theory of fatigue. Trend Eng 13:9–14
10. Paris PC, Erdogan F (1963) A critical analysis of crack propagation laws. ASME J Basic Eng 85D(4):528–533
11. Elber W (1968) Fatigue crack propagation, Ph.D. Thesis, University of New South Wales, Sydney, Australia
12. Elber W (1971) The significance of fatigue crack closure. In: Damage tolerance in aircraft structures, ASTM Special Technical Publication 486, American Society for Testing and Materials, Philadelphia, PA 19103, USA, pp 230–242
13. Hartman A, Schijve J (1970) The effects of environment and load frequency on the crack propagation law for macro fatigue crack growth in aluminium alloys. Eng Fract Mech 1:615–631
14. Chan KS (2004) Variability of large-crack fatigue-crack-growth thresholds in structural alloys. Metall Mat Trans A 35A:3721–3735

© The Author(s), under exclusive licence to Springer Nature B.V., part of Springer Nature 2019   91
R. Wanhill et al., *Fatigue Crack Growth Failure and Lifing Analyses
for Metallic Aircraft Structures and Components*, SpringerBriefs in Applied Sciences
and Technology, https://doi.org/10.1007/978-94-024-1675-6

15. Matsuishi M, Endo T (1968) Fatigue of metals subjected to varying stress – fatigue lives under random loading. In: Preliminary proceedings of the Kyushu District meeting, The Japan Society of Mechanical Engineers, March 1968, pp 37–40

16. Marsh G, Wignall C, Thies PR, Barltrop N, Incecik A, Venugopal V, Johanning L (2016) Review and application of Rainflow residue processing techniques for accurate fatigue damage estimation. Int J Fatigue 82:757–765

17. ASTM E1049-85 (2017) Standard practices for cycle counting in fatigue analysis, ASTM International, West Conshohocken, PA 19428, USA. www.astm.org

18. Schijve J, Jacobs FA, Tromp PJ (1972) Fatigue crack growth in aluminium alloy sheet material under flight simulation loading. Effects of design stress level and loading frequency, NLR Technical Report TR 72018 U, National Aerospace Laboratory NLR, Amsterdam, The Netherlands

19. Barsom JM, (1976) Fatigue crack growth under variable-amplitude loading in various bridge steels. In: Fatigue crack growth under spectrum loads, ASTM Special Technical Publication 595, American Society for Testing and Materials, Philadelphia, PA 19103, USA, pp 217–235

20. Gallagher JP, Stalnaker HD (1978) Developing normalized crack growth curves for tracking damage in aircraft. J Aircraft 15(2):114–120

21. Zhuang WZ, McDonald M, Phillips M, Molent L (2009) Effective block approach for aircraft damage tolerance analyses. J Aircraft 46(5):1660–1666

22. Jones R (2014) Fatigue crack growth and damage tolerance, Fatigue Fract Eng Mater Struct 37:463–483

23. Military Specification Airplane Damage Tolerance Requirements (1974) MIL-A-83444, United States Air Force, The Pentagon, Washington, DC 20330, USA

24. Military Standard Aircraft Structural Integrity Program, Airplane Requirements (1975) MIL- STD-1530A (11), United States Air Force, The Pentagon, Washington, DC 20330, USA

25. Mar JW (1991) Structural integrity of aging airplanes: a perspective. In: Atluri SN, Sampath SG, Tong P (eds) Structural integrity of aging airplanes. Springer, Berlin, Germany, pp 241–262

26. Pearson S (1975) Initiation of fatigue cracks in commercial aluminium alloys and the subsequent propagation of very short cracks. Eng Fract Mech 7:235–247

27. Kitagawa H, Takahashi S (1976) Application of fracture mechanics to very small cracks or the cracks in the early stage. In: Proceedings of the second international conference on mechanical behavior of materials, ASM International, Metals Park, OH 44073, USA, pp 627–631

28. El Haddad MH, Smith KN, Topper TH (1979) Fatigue crack propagation of short cracks, Transactions of the ASME. J Eng Mat Technol 101:42–46

29. Behaviour of Short Cracks in Airframe Components (1983), AGARD Conference Proceedings No. 328, Advisory Group for Aerospace Research and Development, Neuilly-sur-Seine, France

30. Miller KJ, de los Rios ER (eds) (1986) The behaviour of short fatigue cracks, Mechanical Engineering Publications, London, UK

31. Newman JC Jr, Edwards PR (1988) Short-crack growth behaviour in an aluminum alloy— an AGARD cooperative test programme, AGARD Report No. 732, Advisory Group for Aerospace Research and Development, Neuilly-sur-Seine, France

32. Edwards PR, Newman JC Jr (1990) Short-crack growth behaviour in various aircraft materials, AGARD Report No. 767, Advisory Group for Aerospace Research and Development, Neuilly-sur-Seine, France

33. Ritchie RO, Suresh S (1983) Mechanics and physics of the growth of small cracks. In: Behaviour of short cracks in airframe components, AGARD conference proceedings no. 328, Advisory Group for Aerospace Research and Development, Neuilly-sur-Seine, France, pp 1-1–1-14

34. Wanhill RJH (1986) Short cracks in aerospace structures. In: Miller KJ, de los Rios ER (eds) The behaviour of short fatigue cracks. Mechanical Engineering Publications, London, UK, pp 27–36
35. Lincoln JW, Melliere RA (1999) Economic life determination for a military aircraft. J Aircraft 36(5):737–742
36. Jones R, Baker A, Matthews N, Champagne V (eds) (2018) Aircraft sustainment and repair. Butterworth-Heinemann, Elsevier Ltd, Oxford, UK
37. ASTM E647-e1 (2015), Standard test method for measurement of fatigue crack growth rates, ASTM International, West Conshohocken, PA 19428, USA. www.astm.org
38. Hudak SJ Jr, Saxena A, Bucci RJ, Malcolm RC (1978) Development of standard methods of testing and analyzing fatigue crack growth rate data, Technical Report AFML-TR-78-40, Air Force materials Laboratory, Air Force Systems Command, Wright-Patterson Air Force Base, OH 45433, USA
39. Ritchie RO (1977) Near-threshold fatigue crack propagation in ultra-high strength steel: influence of load ratio and cyclic strength. ASME J Eng Mat Technol 99(3):195–204
40. Janssen M, Zuidema J, Wanhill RJH (2002) Fracture mechanics, 2nd edn. Delft University Press, Delft, The Netherlands, pp 211–212
41. Hoeppner DW, Krupp WE (1974) Prediction of component life by application of fatigue crack growth knowledge. Eng Fract Mech 6:47–70
42. Liu HW (1963) Fatigue crack propagation and applied stress range—an energy approach. ASME J Basic Eng 85D(1):116–120
43. Anderson WE (1963) Discussion of Ref [10]. ASME J Basic Eng 85D(4):533
44. Hoeppner DW (2009) Personal communication from the University of Utah, Salt Lake City, UT 84112-9057, USA
45. Forman RG, Kearney VE, Engle RM (1967) Numerical analysis of crack propagation in cyclic-loaded structures. ASME J Basic Eng 89D(3):459–463
46. Newman JC Jr (1984) A crack opening stress equation for fatigue crack growth. Int J Fract 24:R131–R135
47. Maierhofer J, Pippan R, Gänser H-P (2014) Modified NASGRO equation for physically short cracks. Int J Fatigue 59:200–207
48. Forman RG, Mettu SR, (1992) Behavior of surface and corner cracks subjected to tensile and bending loads in Ti-6Al-4V alloy. In: Ernst HA, Saxena A, McDowell DL (eds) Fracture mechanics, 22nd symposium, ASTM Special Technical Publication 1131, vol 1, American Society for Testing and Materials, Philadelphia, PA 19103, USA, pp 519–546
49. Fatigue Crack Growth Computer Program "NASA/FLAGRO" (1989) Users' Manual, JSC-22267, NASA Lyndon B. Johnson Space Center, Houston, TX77058, USA
50. Hu W, Tong YC, Walker KF, Mongru D, Amaratunga R, Jackson P (2006) A review and assessment of current lifing methodologies and tools in Air Vehicles Division, DSTO Research Report DSTO-RR-0321, DSTO Defence Science and Technology Organisation, Fishermans Bend, Victoria 3207, Australia
51. Iyyer N, Sarkar S, Merrill R, Phan N (2007) Aircraft life management using crack initiation and crack growth models—P-3C aircraft experience. Int J Fatigue 29:1584–1607
52. Schijve J (2009) Fatigue of structures and materials, 2nd edn. Springer Science + Business Media, Dordrecht, The Netherlands, pp 351–361
53. Newman JC Jr (1992) FASTRAN-II—a fatigue crack growth structural analysis program, NASA Technical Memorandum 104159, National Aeronautics and Space Administration, Langley Research Center, Hampton, VA 23665, USA
54. Paris PC, Tada H, Donald JK (1999) Service load fatigue damage—a historical perspective. Int J Fatigue 21:S35–S46
55. Ciavarella M, Paggi M, Carpinteri A (2008) One, no one, and one hundred thousand crack propagation laws: a generalized Barenblatt and Botvina dimensional analysis approach to fatigue crack growth. J Mech Phys Solids 56:3416–3432

56. Irving PE, Lin J, Bristow JW (2003) Damage tolerance in helicopters: report on the Round Robin challenge. In: 59th American Helicopter Society International Annual Forum 2003, Phoenix, AZ 85003, USA, vol 3, Curran Associates Inc., Red Hook, NY 12571, USA, pp 1642–1652

57. Vaughan RE, Chang JH (2003) Life predictions for high cycle dynamic components using damage tolerance and small threshold cracks. In: 59th American Helicopter Society International Annual Forum 2003, Phoenix, AZ 85003, USA, vol 3, Curran Associates Inc., Red Hook, NY 12571, USA, pp 1712–1720

58. Tiong U, Jones R (2009) Damage tolerance analysis of a helicopter component. Int J Fatigue 31(6):1046–1053

59. Wanhill RJH, Molent L, Barter SA (2013) Fracture mechanics in aircraft failure analysis: uses and limitations. Eng Failure Anal 35:33–45

60. Gallagher JP, Miedlar PC, Juarez V (1990) Rapid repair DTA technology for F-16 aircraft. In: Cooper TD, Lincoln JW (eds) Proceedings of the 1989 Structural Integrity Program Conference, Technical Report WRDC-TR-90-4051, Wright-Patterson Air Force Base, OH 45433, USA, pp 497–518

61. Wanhill RJH (1991) Durability analysis using short and long fatigue crack growth data. In: Jones R, Miller NJ (eds) International conference on aircraft damage assessment and repair. Institution of Engineers, Barton, ACT, Australia, pp 100–104

62. De Jonge JB (1994) The crack severity index of monitored load spectra. In: An assessment of fatigue damage and crack growth prediction techniques, AGARD Report 797, Advisory Group for Aerospace Research and Development, Neuilly-sur-Seine, France, pp 5-1–5-5

63. Wanhill RJH (1994) Damage tolerance engineering property evaluations of aerospace aluminium alloys with emphasis on fatigue crack growth, NLR Technical Publication TP 94177 U. National Aerospace Laboratory NLR, Amsterdam, The Netherlands

64. McDonald M (2013) Guide on the Effective Block Approach for fatigue life assessment of metallic structures, DSTO Technical Report DSTO-TR-2850, DSTO Defence Science and Technology Organisation, Fishermans Bend, Victoria 3207, Australia

65. Molent L, Barter SA, Wanhill RJH (2010) The lead crack fatigue lifing framework, DSTO Research Report DSTO-RR-0353, DSTO Defence Science and Technology Organisation, Fishermans Bend, Victoria 3207, Australia

66. Molent L, Barter SA, Wanhill RJH (2011) The lead crack fatigue lifing framework. Int J Fatigue 33:323–331

67. Barter S, McDonald M, Molent L (2005) Fleet fatigue life interpretation from full-scale and coupon fatigue tests—a simplified approach. In: Proceedings, 2005 USAF structural integrity program (ASIP) conference, 29 November–1 December 2005, Memphis, TN 38103, USA: Available on CD-ROM from ASIP 2005 c/o Universal Technology Corporation 1270 North Fairfield Road Dayton, OH 45432-2600, USA

68. Molent L, McDonald M, Barter S, Jones R (2008) Evaluation of spectrum fatigue crack growth using variable amplitude data. Int J Fatigue 30:119–137

69. Gallagher JP (1976) Estimating fatigue-crack lives for aircraft: techniques. Exp Mech 16 (11):425–433

70. Barter SA, Wanhill RJH (2008) Marker loads for quantitative fractography (QF) of fatigue in aerospace alloys, NLR Technical Report NLR-TR-2008-644. National Aerospace Laboratory NLR, Amsterdam, The Netherlands

71. Eijkhout MT (1994) Fractographic analysis of longitudinal fuselage lapjoint at stringer 42 of Fokker 100 full scale test article TA15 after 126350 simulated flights, Fokker Report RT2160. Fokker Aircraft Ltd., Amsterdam, The Netherlands

72. Wanhill RJH, Hattenberg T (2006) Fractography-based estimation of fatigue "initiation" and growth lives in aircraft components. In: Proceedings of the international conference on structural integrity and failure (SIF 2006), 27–29 September 2006, Sydney, Australia, Institute of Materials Engineering Australasia Ltd, Melbourne, Australia, pp 196–209

73. Wanhill RJH, Symonds N, Merati A, Pasang T, Lynch SP (2013) Five helicopter accidents with evidence of material and/or design deficiencies. Eng Failure Anal 35:123–146

74. Speaker SM, Gordon DE, Kaarlela WT, Meder A, Nay RO, Nordquist FC, Manning SD (1982) Durability methods development, Volume VIII—Test and fractography data, AFFDL Technical Report AFFDL-TR-79-3118, USAF Air Force Flight Dynamics Laboratory, Wright-Patterson Air Force Base, OH 45433, USA

75. Berens AP, Hovey PW, Skinn DA (1991) Risk analysis for aging aircraft fleets, Volume 1 —Analysis, Wright Laboratory Technical Report WL-TR-91-3066, USAF Flight Dynamics Directorate (WL/FIBEC), Wright-Patterson Air Force Base, OH 45433-6553, USA

76. Goldsmith NT, Clark G, Barter SA (1996) A growth model for catastrophic cracking in an RAAF aircraft. Eng Failure Anal 3(5):191–201

77. Molent L, Singh R, Woolsey J (2005) A method for evaluation of in-service fatigue cracks. Eng Fail Analysis 12(1):13–24

78. Barter S, Molent L, Goldsmith N, Jones R (2005) An experimental evaluation of fatigue crack growth. Eng Fail Anal 12(1):99–128

79. Molent L, Barter SA (2007) A comparison of crack growth behaviour in several full-scale airframe fatigue tests. Int J Fatigue 29(6):1090–1099

80. Underhill PR, DuQuesnay DL (2008) The effect of dynamic loading on the fatigue scatter factor for Al 7050. Int J Fatigue 30(4):614–622

81. Liao M, Benak T, Renaud G, Yanishevsky M, Marincak T, Bellinger NC, Mills T, Prost-Domansky S, Honeycutt K (2008) Development of short/small crack model for airframe material: 7050 aluminum alloys. In: The 11th Joint NASA/FAA/DOD Conference on Aging Aircraft, 21-24 April, 2008, Phoenix, AZ 85003, USA: Available on CD-ROM from Washington, D.C.: Federal Aviation Administration: Department of Defense: National Aeronautics and Space Administration

82. Mohanty JR, Verma BB, Ray PK (2009) Prediction of fatigue life with interspersed mode-I and mixed-mode (I and II) overloads by an exponential model: extensions and improvements. Eng Fract Mech 76(3):454–468

83. Royal Australian Air Force (2007) Structural analysis methodology—F/A-18A/B, ASI/2006/1114755 Pt 1 (18), Issue 2, AL2, 7 December 2007. Canberra, Australia

84. Military Specification Airplane Damage Tolerance Requirements (1974) MIL-A-83444, United States Air Force, The Pentagon, Washington, D.C. 20330, USA

85. Molent L, Barter SA, Dixon B, Swanton G (2018) Outcomes from the fatigue testing of seventeen fuselage structures. Int J Fatigue 111:220–232

86. Molent L (2014) A review of equivalent pre-crack sizes in aluminium alloy 7050-T7451. Fatigue Fract Eng Mat Struct 37:1055–1074

87. Gallagher JP, Molent L (2015) The equivalence of EPS and EIFS based on the same crack growth life data. Int J Fatigue 80:162–170

88. Boykett R, Walker K, Molent L (2008) Sole operator support for the RAAF F-111 fleet. In: The 11th Joint NASA/FAA/DOD Conference on Aging Aircraft, 21–24 April, 2008, Phoenix, AZ 85003, USA: Available on CD-ROM from Washington, D.C.: Federal Aviation Administration: Department of Defense: National Aeronautics and Space Administration

89. Clark G, Jost GS, Young GD (1997) Recovery of the RAAF Macchi MB326H—the tale of an ageing trainer fleet. In: Cook R, Poole P (eds) Fatigue in new and ageing aircraft, Proceedings of the 19th ICAF symposium, EMAS, Warley, UK, pp 39–58

90. Wanhill R, Molent L, Barter S (2016) Milestone case histories in aircraft structural integrity. In: Hashmi S (ed) Reference module in materials science and materials engineering, Elsevier Inc., Oxford, UK

91. Molent L, Jones R (2016) A stress versus crack growth rate investigation (aka stress–cubed rule). Int J Fatigue 87:435–443

92. Ayling J, Bowler G, Brick M, Ignjatovic M (2014) Practical application of structural repair fatigue life determination on the AP-3C Orion platform. Adv Mat Res 891–892:1065–1070

93. Wanhill RJH, de Graaf EAB, van der Vet WJ (1974) Investigation into the cause of an S-61 N helicopter accident. Part I: Fractographic analysis and blade material tests, NLR Technical Report NLR TR 74103 C, National Aerospace Laboratory NLR, Amsterdam, The Netherlands

94. Hattenberg T, Vlieger H (1993) Fractographic investigation of cracks in finger strip (LHS) and rib 5.0 (RHS) of fin found during the 2nd phase of Fokker 100 TA-22B testing, NLR Contract Report NLR CR 93265 C, National Aerospace Laboratory NLR, Amsterdam, The Netherlands

95. Wanhill RJH, Koolloos MFJ (2001) Fatigue and corrosion in aircraft pressure cabin lap splices. Int J Fatigue 23:S337–S347

96. Anderson WE (1972) Fatigue of aircraft structures. Int Metall Rev 17:240–263

97. AGARD Report No. 659 (1977) Corrosion fatigue of aircraft materials, Advisory Group for Aerospace Research and Development, Neuilly-sur-Seine, France

98. Barter S, Sharp PK, Clark G (1994) The failure of an F/A-18 trailing edge flap hinge. Eng Fail Anal 1(4):255–266

99. Byrnes R, Goldsmith N, Knop M, Lynch S (2014) Corrosion-fatigue crack growth in age-hardened Al alloys. Adv Mat Res 891–892:248–253

100. Trathan P (2011) Corrosion monitoring systems on military aircraft. In: Proceedings of the 18th international corrosion conference, 20–24 November 2011, Perth, Australia, vol 2: Available on CD-ROM from Curran Associates Inc., Red Hook, NY 12571, USA, pp 1231–1240

101. Barter SA, Molent L (2013) Service fatigue cracking in an aircraft bulkhead exposed to a corrosive environment. Eng Fail Anal 34:181–188

102. Molent L (2015) Managing airframe fatigue from corrosion pits—a proposal. Eng Fract Mech 137:12–25

103. Brooks CR, Choudhury A (2002) Failure analysis of engineering materials, The McGraw-Hill Education, New York, NY 101221, USA

104. Findley SJ, Harrison ND (2002) Why aircraft fail. Mat Today 5(11):18–25

105. Campbell GS, Lahey R (1984) A survey of serious aircraft accidents involving fatigue fracture. Int J Fatigue 6(1):25–30

106. Wanhill RJH (2009) Some notable aircraft service failures investigated by the National Aerospace Laboratory (NLR). Struct Integrity Life 9(2):71–87

107. Tiffany CF, Gallagher JP, Babish CA IV (2010) Threats to structural safety, including a compendium of selected structural accidents/incidents, USAF Technical Report ASC-TR-2010-5002, Aeronautical Systems Center Engineering Directorate, Wright-Patterson Air Force Base, OH 45433-7101, USA

108. Barter SA, Molent L, Wanhill RJH (2012) Typical fatigue-initiating discontinuities in metallic aircraft structures. Int J Fatigue 41:11–22

109. Fenner AJ, Field FE (1960) A study of the onset of fatigue damage due to fretting, North East Coast Institution of Engineers and Shipbuilders Transactions 76:183–228

110. Edwards PR (1981) The application of fracture mechanics to predicting fretting fatigue. In: Waterhouse RB (ed) Fretting fatigue. Applied Science Publishers Ltd, London, UK, pp 67–97

111. Wanhill RJH, Platenkamp DJ, Hattenberg T, Bosch AF, de Haan PH (2009) GLARE teardowns from the MegaLiner Barrel (MLB) fatigue test. In: Bos M (ed) ICAF 2009, Bridging the gap between theory and operational practice. Springer Science + Business Media, The Netherlands, Dordrecht, pp 145–167

112. Wanhill RJH (2018) Discussions and comments on the paper E.S. Dzidowski, 2013, "The effect of secondary metalworking processes on susceptibility of aircraft to catastrophic failures and prevention methods, Archives of Metallurgy and Materials, 58(4), pp 1207–1212", Arch Metall Mat 63(2), 773–779

113. Davidson D, Chan K, McClung R, Hudak S (2003) Small fatigue cracks. In: Ritchie RO, Murakami Y (eds) Comprehensive structural integrity, vol 4: Cyclic loading and fatigue. Elsevier Science, Amsterdam, The Netherlands, pp 129–164

114. Molent L (2010) Fatigue crack growth from flaws in combat aircraft. Int J Fatigue 32:639–649

115. Komorowski JP, Bellinger NC, Liao M, Fillion A (2007) Application of the holistic structural integrity process to Canadian Forces challenges. In: Proceedings, 2007 USAF structural integrity program (ASIP) conference, 4–6 Dec 2007, Palm Springs, CA 92262, USA: www.asipcon.com/pages/proceedings.html

116. Lynch SP (2017) Some fractographic contributions to understanding fatigue crack growth. Int J Fatigue 104:12–26

117. Gough HJ (1926) The Fatigue of Metals. Ernest Benn Ltd., London, UK

118. McEvily AJ, Matsunaga H (2010) On fatigue striations, Scientia Iranica Transactions B. Mech Eng 17(1):75–82

119. Wanhill RJH (2003) Material-based failure analysis of a helicopter rotor hub. Pract Fail Anal 3(2):59–69

120. Forsyth PJE (1976) The causes of mixed fatigue/tensile crack growth and the significance of microscopic crack behaviour, RAE Technical Report TR 75143, February 1976. Royal Aircraft Establishment, Farnborough, UK

121. Vlasveld JA, Schijve J (1980) Tongue-shaped crack extension during fatigue of high strength aluminium alloys. Fatigue Eng Mat Struct 3(2):129–145

122. Yoder GR, Cooley LA, Crooker TW (1979) 50-fold difference in region-II fatigue crack propagation resistance of titanium alloys: a grain size effect. Trans Am Soc Mech Eng, J Eng Mat Technol 101(1):86–90

123. Goldsmith NT (1978) Fractographic examinations relevant to the F + W Mirage fatigue test, DSTO Technical Memorandum ARL-MAT-TM-371, DSTO Defence Science and Technology Organisation, Fishermans Bend, Victoria 3207, Australia

124. Cox AF, Barter SA (2003) F/A-18 horizontal stabilator spindle AF1410 coupon testing programme, DSTO Technical Report DSTO-TR-1443, DSTO Defence Science and Technology Organisation, Fishermans Bend, Victoria 3207, Australia

125. Goldsmith NT (2000) Deep focus: a digital image processing technique to produce improved focal depth in light microscopy. Image Anal Stereol 19:163–167

126. Swain MH, Newman JC Jr (1984) On the use of marker loads and replicas for measuring growth rates for small cracks. In: Fatigue crack topography, AGARD conference proceedings No. 376, Advisory Group for Aerospace Research and Development, Neuilly- sur-Seine, France, pp 12-1–12-17

127. Wanhill RJH, Schra L (1990) Short and long fatigue crack growth in 2024-T3 under Fokker 100 spectrum loading. In: Edwards PR, Newman JC Jr (eds) Short-crack growth behaviour in various aircraft materials, AGARD Report No. 767, Advisory Group for Aerospace Research and Development, Neuilly-sur-Seine, France, pp 8-1–8-26

128. Merati A (2005) A study of nucleation and fatigue behaviour of an aerospace aluminum alloy 2024-T3. Int J Fatigue 27:33–44

129. Merati A, Eastaugh G (2007) Determination of fatigue related discontinuity state of 7000 series of aerospace aluminum alloys. Eng Fail Anal 14:673–685

130. Barter S, Russell D (2013) Corrosion pitting as a fatigue crack initiator in AA7050-T74511 under a fighter spectrum, DSTO Technical Report DSTO-TR-2913, DSTO Defence Science and Technology Organisation, Fishermans Bend, Victoria 3207, Australia

131. Murakami Y (2002) Metal fatigue: effects of small defects and nonmetallic inclusions. Elsevier Science Ltd., Oxford, UK

132. Raju IS, Newman JC Jr (1979) Stress intensity factors for a wide range of semi-elliptical surface cracks in finite thickness plates. Eng Fract Mech 11(4):817–829

133. Barter SA, Clark G, Goldsmith NT (1993) Influence of initial defect conditions on structural fatigue in RAAF aircraft. In: Blom AF (ed) ICAF '93, durability and structural integrity of airframes, vol 1. EMAS. Warley, UK, pp 281–304

134. Aviation Safety Investigation Report 199400519 (1994) Fokker B.V. F50 (Fokker 50), Australian Transport Safety Bureau, Canberra, ACT 2601, Australia

135. Zhuang W, Molent L (2008) Block-by-block approaches for spectrum fatigue crack growth prediction. Eng Fract Mech 75(17):4933–4947

136. Molent L, Barter SA, White P, Dixon B (2009) Damage tolerance demonstration testing for the Australian F/A-18. Int J Fatigue 31:1031–1038

137. McDonald M, Molent L (2004) Fatigue assessment of the F/A-18 horizontal stabilator spindle, DSTO Technical Report DSTO-TR-1620, DSTO Defence Science and Technology Organisation, Fishermans Bend, Victoria 3207, Australia

138. Barter SA, Molent L, (2012) Investigation of an in-service crack subjected to aerodynamic buffet and manoeuvre loads and exposed to a corrosive environment. Paper ICAS 2012-7.9.1, 28th congress of the international council of the aeronautical sciences, 23–28 September 2012, Brisbane, Australia

139. Potter JM, Gallagher JP, Stalnaker HD (1974) The effect of spectrum variations on the fatigue behavior of notched structures representing F-4E/S wing stations, USAF Technical Memorandum AFFDL-TM-74-2-FBR, Air Force Flight Dynamics Laboratory Director of Laboratories, Air Force Systems Command, Wright-Patterson Air Force Base, OH 45433, USA

140. Jongebreur AA, Louwaard EP, Van der Velden RV (1985) Damage tolerance test program of the Fokker 100. In: Salvetti A, Cavallini G (eds) Durability and damage tolerance in aircraft design, Proceedings of the 13th ICAF symposium, EMAS, Warley, UK, pp 317-349

141. Wanhill RJH, De Graaf EAB, Delil AAM (1979) Significance of a rotor blade failure for fleet operation, inspection, maintenance, design and certification, Paper Nr. 38. In: Proceedings of the fifth European rotorcraft and powered lift aircraft forum, 4–7 Sept 1979, vol 1, National Aerospace laboratory NLR, Amsterdam, The Netherlands

142. Rich MJ (1971) Crack propagation in helicopter rotor blades. In: Damage tolerance in aircraft structures, ASTM Special Technical Publication 486, American Society for Testing and Materials, Philadelphia, PA 19103, USA, pp 243–251

143. Cook EB (1961-4) Replacement time substantiation of the S-61L and S61-R (CH-3C) main rotor blade based on laboratory fatigue tests, Sikorsky Aircraft Report SER-61368, Sikorsky Aircraft Corporation, Stratford, CT 06614-1385, USA

144. Falconer JM, Goddard PN (1998) GKN Westland Helicopters letter JMF/KP/1036 to the Royal Netherlands Navy, 14 Nov 1998

145. GKN Westland, KIM (RNLN) and NLR (2000) Material-based failure analysis of the Lynx-282 rotor hub, NLR Contract Report NLR-CR-99189C, National Aerospace Laboratory NLR, Amsterdam, The Netherlands

146. Wanhill RJH, Looije CEW (1993) Fractographic and microstructural analysis of fatigue crack growth in Ti-6Al-4V fan disc forgings. In: AGARD engine disc cooperative test programme, AGARD Report 766 (Addendum), Advisory Group for Aerospace Research and Development, Neuilly-sur-Seine, France, pp 2-1–2-40

147. Murakami Y (1987) Stress intensity factors handbook, vol 2. Pergamon Press, New York, NY 10523, USA, p 661

148. Bannantine JA, Comer JJ, Handrock JL (1990) Fundamentals of Metal Fatigue Analysis. Prentice-Hall, Inc., New Jersey, NJ 07632, USA, p 95

149. Laméris J (2000) Review of the fatigue analysis of the RNLN Lynx main rotor head, NLR Contract Report NLR-CR-2000-193C, National Aerospace Laboratory NLR, Amsterdam, The Netherlands

150. King SP (1999) Lynx hub MPOG fatigue damage assessment, Report CD/SPK/JTWP20-3555, Issue 1. GKN Westland Helicopters Limited, Yeovil, UK
151. Terlinde G, Fischer G (2003) Beta titanium alloys. In: Leyens C, Peters M (eds) Titanium and titanium alloys. Wiley-VCH Verlag GmbH & Co. KGaA, Weinheim, Germany, pp 37–57

Printed in the United States
By Bookmasters